S0-ABM-689

Herman the German

Herman the German

Enemy Alien
U.S. Army
Master Sergeant
#10500000

by Gerhard Neumann

William Morrow and Company, Inc.
New York 1984

Library of Congress Cataloging in Publication Data

Neumann, Gerhard, 1917-
Herman the German.

1. Neumann, Gerhard, 1917- . 2. Mechanical engineers—United States—Biography. I. Title.
TJ140.N35A35 1984 621'.092'4 83-23708
ISBN 0-688-01682-0

Printed in the United States of America

3 4 5 6 7 8 9 10

BOOK DESIGN BY JAMES UDELL

FOREWORD

The first time I met Gerhard Neumann was in Kunming, China, during World War II. Ever since, I have been fascinated by the kaleidoscopic adventures crowding his life. After meeting my late husband, General Claire Lee Chennault, Gerhard joined the famous Flying Tigers, where he was known as "Herman the German." Paradoxically, although an enemy alien, Herman the German became a master sergeant in the U.S. Army Air Corps.

Gerhard's engineering talents, his zest, his sense of humor, and his magnificent rapport with people at all levels won him the respect and affection of everyone who met him. General Bruce Holloway, former head of the Strategic Air Command and vice-chief of our Air Force, once said: "He was a gigantic asset to the operations over there in China. Having Gerhard as a line chief was like having Charles Kettering [General Motors' famous engineer and inventor] run the local Chevrolet maintenance shop."

Neumann's phenomenal success in assembling a Japanese Zero fighter plane and his work with the OSS in the Orient earned him his U.S. citizenship by act of Congress. After the war, a 10,000-mile Jeep trip across Asia with his American wife and dog, and an astonishing career as a maverick-type manager—he rose to head up General Electric's multibillion-dollar jet engine business—added more adventure and achievement to his amazing life.

Gerhard Neumann represents that extraordinary blend of varied heritages, cultures and talents that has made America a country not only of the past but definitely of the future.

Anna Chennault

by Anna Chennault

ACKNOWLEDGMENTS

My sincere thanks to Anna Chennault and Bill Schoneberger who urged, cajoled and inspired me three years ago to undertake writing this autobiography. And where would this book be without the confidence of Howard Cady?

I also want to thank and express my deep appreciation to my good friends Joanne Grebe, Shirley Clarke, Bob Hotz, Bill Schoneberger (again!) and my son Lee Duane who patiently plowed through a rough draft of my manuscript and who, by their constructive comments, encouraged me to finish the book. Toni Valk did an amazing job of checking facts and correcting my spelling and punctuation.

Finally, my gratitude to my wife, Clarice, who not only participated in many of the adventures, but also in the preparation of this final manuscript.

—GERHARD NEUMANN
Swampscott, Massachusetts

CONTENTS

1

A Split Second Made the Difference

June 27, 1940. It was that hot, muggy part of the day when Hong Kong commuters between Victoria and the Kowloon peninsula (on the Chinese mainland) really appreciated the open-sided, breezy STAR ferries crisscrossing the picturesque harbor crowded with British warships, freighters and large sailing junks. "Invincible" France had just surrendered to Hitler's Germany, following a twenty-two-day surprise blitzkrieg. English troops, too, had taken a terrible shellacking at Dunkirk and lost all of their equipment, and most were fortunate to be able to escape back to England. Here, in the most Far Eastern base of Great Britain, a British Tommy carrying his heavy Enfield rifle with fixed bayonet was walking three steps behind me and to the left—as he had been doing since nine o'clock that morning and all of the day before. We crossed Connaught Road with its double-decker green streetcars and red Thornycroft buses, leaving the shaded sidewalk of the Holland-America Bank to get to the only modern high-rise building in the British Crown Colony, the attractive twenty-story Hong Kong & Shanghai Bank.

This was my last stop after seven frustrating visits to representatives of foreign countries and businesses before returning to the barbed-wire-encircled LaSalle College in Kowloon for the interned

Germans. Two days earlier, the British camp commander, for a reason we could never understand except that he wanted to get rid of as many of us as possible, gave each "enemy alien"—there were 102 of us—forty-eight hours in which to secure an entry permit into any country with which Hitler was not yet at war. The colonel's generous-sounding but unrealistic proposal was for us to "get out of Hong Kong" by midnight of the second day or be shipped to a newly established internment camp on the island of Ceylon in the Bay of Bengal "for the duration" of the war, however long this might turn out to be. The fact that our British captors kept in their possession both our German travel and military passports (taken away from us as we were individually arrested during the night of September 3, 1939, when England declared war on Germany) made success in getting out of Hong Kong most unlikely. So much so that only a handful of internees even attempted to leave the Crown Colony. After my fruitless visits to five consulates and the representatives of two American automobile assembly plants in the Philippines, I had picked the in-town office of CNAC (Chinese National Aviation Corporation) as a last possibility to get away from the British. I had not the slightest idea how this could be accomplished.

CNAC's office was on the eighth floor of the gleaming white building. The young receptionist, sleek in her typical black Chinese gown slit on the side way above her knee, told me, "Boss out. You come back tomorrow!"

Disheartened, I said to my guard, "Let's go back to camp."

I pressed the down button and watched the little red floor numbers flash by as one of the four elevators came up to the eighth floor to fetch us. The door opened. I stepped into the car closely followed by my guard—and bumped into a tall, middle-aged gentleman who was just going to exit at the floor we were on. I stepped back, stammering my apologies in German and English for not letting him get off first. The man, hearing my accent and seeing the armed soldier, was startled. "Wait a minute! What is going on here?" By that time, the elevator door had shut automatically. The car was on its way down without us.

It was an incredible coincidence of split-second timing . . . my running into Mr. W. Langhorne Bond, vice president of the Pan American Airways Pacific Division, who happened to be visiting

CNAC's office while on a brief turnaround of the China Clipper flying boat which commuted from San Francisco to Hong Kong via Manila twice weekly. Sitting on the edge of a desk, the American questioned me about my situation, then picked up the phone and made the phone call that gave me my ticket to freedom.

And that's the way much of my exciting life has gone: a series of coincidences of meeting the right people, at the right place, at the right time.

Now, sixty-six years old, a retired General Electric executive struggling three years to write this autobiography but enjoying the effort, I appreciate the impact my German upbringing—the years of discipline at home and at school, the tough "dirty fingernail" apprenticeship and the *practical* college engineering education—had on my career. Some people feel that the writer of an autobiography must be out to make an ego trip. I can assure you that this certainly is not my intent. I want the reader to share in my unique adventures, and to benefit from the lessons I learned.

Remember the attractive redhead, Miss Christine Keeler, who caused British Secretary of Defense John Profumo to resign and a severe British Cabinet crisis in the early sixties? When asked by a reporter how she became such a successful prostitute, her now-famous reply was "Just lucky, I guess. . . ." I liked Miss Keeler's reply. Being lucky, many times, was a most important ingredient in my life.

2

Apprenticeship in Germany: No Bed of Roses!

"Understand, *Junge* [boy]?" is still ringing in my ears. For the first two years of my three-year apprenticeship in Frankfurt an der Oder near the Polish border in Germany, Meister Schroth asked me repeatedly this question. Looking back on the years spent with this man, I realize I learned to work and to be precise; not to be concerned about "honestly" acquired black fingernails but rather to be proud of them. As a young man I had not the slightest idea that Herr Schroth's "Do it right!" attitude, his insistence on perfection and his confidence in his ability to handle any repair on anything mechanical would shape my career: in China, in Afghanistan, in America, in Viet Nam . . . in war and in peace; working on German, British and American cars and airplanes; assembling the first captured Japanese Zero fighter aircraft in 1942, a nemesis to our pilots and a technical mystery at that time; making one Jeep out of two damaged ones that would carry my wife, myself and our Airedale terrier 10,000 miles across Asia in 1947; designing aircraft engines at General Electric or patenting ideas which are now used worldwide; advising the U.S. Army on the engine for its latest sixty-ton heavy M-1 tank. . . . There was never a time in my life that the training by Herr Schroth did not affect me.

In early 1933 I passed, but barely, my Gymnasium graduation

examinations. Three foreign languages plus math, chemistry, physics and geography were supposedly firmly implanted in my head. I had always been interested in mechanical gadgetry, watching our chauffeur (a World War I buddy of my father's) in the small workshop he maintained in the garage of our big Mercedes. I was fortunate to get from my parents mechanical building sets as Christmas and birthday presents. The decision not to join my father in his feather business but instead to become an engineer had been made: I wanted to invent something, watch it being made and see it being used. Two years of machinist or mechanic apprenticeship were mandatory for admission to any German engineering academy or college. Although it was not essential, many parents desired to sign up their sons for a three-year apprenticeship and then have them pass a lengthy series of tests given by the craft guild. So it was with my father: He wanted me to be a licensed journeyman prior to my embarking on an engineering profession.

No one lower than a craft *Meister* was permitted, under the law, to employ apprentices and train them. My dad asked Frankfurt/Oder's ten cabdrivers to recommend a garage where they thought I would learn most. Their unanimous reply: at Alfred Schroth's. Few automobile owners in town had ever heard of Schroth, Frankfurt's smallest garage. But professional cabdrivers knew him well. No wonder: He was a master mechanic, expert welder and *Meister in Maschienenbau* (construction of machine tools) who wanted no help from journeymen, and certainly not from apprentices. Schroth may have been the originator of the axiom "If you want something done right, do it yourself." As a result of the cabbies' recommendation and alerted to Schroth's attitude, my father took on the challenge of trying to convince Herr Schroth to accept me as his first apprentice for three full years. During the hour-long discussion between my father and Schroth outside his garage entrance, I stood mute. I did not say one word—nor was I asked a single question. Finally, Schroth agreed to take me.

He had "neither time nor the inclination," I heard Schroth tell my father, to do things twice. He was thorough in his work—and he "gave a damn." He was a South German, a man of few words. He was married, without children and always without money. Schroth didn't care how much or how little he earned as long as he had enough to provide for his wife and his huge St. Bernard (who slept in the garage at night and left a greeting which I had to clean up "first thing every morning!"). Schroth was proud of his work. His self-satisfaction as a fine craftsman was worth everything to him —far more than money. Several times I saw him send some fellow on his way after fixing his car, without asking for any payment.

I will always remember my first day of the apprenticeship. Meister Schroth looked at my brand-new, still-stiff blue coveralls, then at my clean white hands. "*Junge,* I want everything done right here, as it was when you were *not* around! How much you learn is up to you. One more thing: I shall never say thank you. Is that clear, boy?" Indeed it was. I was not quite sixteen years old. In three subsequent years of hard work—and I do mean hard work—there was never a thank-you from him or a single word of encouragement or approval.

The apprenticeship period, formally contracted between *Meister* and father, and approved by the guild, had several minuses and pluses for the employer. On one hand, a master was contractually obligated to do his best to teach his apprentice the trade; he also would have to support him with food, clothing and housing if the boy's parents were unable to do so. If the apprentice quit prematurely, however, his parents had to pay a penalty. On the plus side, not only was there an extra pair of hands available to him, but an apprentice received no pay during his first two years of training, was not entitled to any vacation, and had to work as many hours as the boss wanted him to. During his third year, an apprentice received pocket money for streetcar fares, gas for his motorcycle or for a couple of movie tickets. At first, I had second thoughts about the wisdom of my having become a mechanic's apprentice: My friends were still continuing at the Gymnasium, spending their days in comfortable and clean surroundings; here I was, accustomed to a fine home and the luxury of two maids and chauffeur, becoming a grease monkey for three long years. I didn't approach the prospect with enthusiasm.

Schroth's standards were tough. He gave me his basic rules within minutes after I entered the garage (I bet he thought about them all night long!):

—"Never lean against the workbench. If you are too tired to stand up straight, go home. Understand, boy?

—"All tools must be cleaned and placed in their proper drawers or hung on the proper hooks before you go home at night. Understand, boy?

—"When you file or saw metal, don't move the tool like a virtuoso strokes his violin with his bow. Move the saw or file straight and level. Understand, boy?

—"All slots of screws are to be turned to the same direction. If a bolt is too loose or too tight as a consequence, use a different-thickness washer, or grind the one you have. Understand, boy?

—"When I tell you to clean a part, I want it clean. Understand, boy?"

And a few more rules, simple and clear.

—"Yes, master," "Yes, master," "Yes, master" was my only response.

Meister Schroth wore a green shirt, a black bow tie, blue work pants held up by black suspenders and a blue Eisenhower jacket every day the three years I was with him. His garage was simple; it was barely big enough for two cars, one behind the other. His tow truck was a Harley-Davidson motorcycle with sidecar. His garage had neither a hydraulic hoist nor an air compressor. Patched tires had to be pumped by hand; he had not even a coaster-on-wheels on which a mechanic could slide on his back underneath a vehicle for transmission or clutch repairs, to adjust brakes or to change engine oil. The Schroth method was a potato sack under one's back. When I wormed myself underneath a muddy car in winter to adjust brake linkages, dirt or snow dropped from the chassis—only inches above the tip of my nose—into eyes and mouth. It quickly was good-bye to clean hands and clean fingernails. Oil, grease, rusty wires, slipping tools, bent cotter pins, sharp corners of cracked fenders—everything took its toll on hands and knuckles.

In winter my hands were frozen purple. Wear work gloves? "What's the matter, boy, are you a girl?" When my hands were bleeding, Herr Schroth pointed to the large bottle of iodine in the backroom and mumbled something about *faules Fleisch* (lazy flesh). No Band-Aids, no pitying, no time out. One day, as a result of my carelessness, a fifty-pound motorcycle engine toppled off the workbench, heading for the concrete floor; the spark plug's threaded end protruded from the motor like a fuse of a bomb. Instinctively, I jumped forward and thrust my hand under the engine to try to break its fall. The plug's threaded end shot through my right palm, sticking out cleanly between the first and second finger bones (there is still a scar today). Even then there was no sympathy from Herr Schroth, although he *did* help pull my impaled hand off the plug. Fortunately for me—and to Schroth's relief—I had saved the engine; only the plug had to be replaced. He gave me hell anyhow.

Getting grease and dirt off hands each noon and evening was a new experience. Schroth showed me how. We brushed our hands in gasoline kept in an open chamber pot, dipped them in a pail of used engine oil, gathered a handful of sawdust out of a paper bag and rubbed the hands vigorously with the mixture. Incredibly, the color of our hands changed from oily black to "clean" gray. Finally, a good hand-wringing with water and soap: My hands were clean enough to reach into the pocket of my overalls to get streetcar fare home or to ride the bike. On rare occasions when I had a date, I literally sat on my hands because I was embarrassed by the black under my fingernails, which would not get clean regardless of how hard I scrubbed them.

To Schroth it didn't matter how late beyond 6 P.M. we worked. Soon it didn't matter to me, either. I actually began to welcome the long hours of effort. We did motor- and bodywork, from engine overhauls to tune-ups, from straightening fenders to painting whole bodies. I learned to fix electrical systems, to gas-weld with oxygen and acetylene, and also to fix fenders with an electric arc welder. The master showed me how to machine parts on our lathe. There was nothing on an automobile or truck which Herr Schroth did not tackle—and I with him—even if it meant that I only cleaned parts or held the spotlight steady so that he could see better. (To keep a hand lamp steady for fifteen minutes—so that its beam shed light precisely on the spot where the *Meister* worked—was not an easy matter.)

During my first year of apprenticeship, I acted more like a dentist's nurse: I cleaned engine parts, handed Schroth the proper tools, prepared the patient for the doctor's work. . . . We worked on motorcycles and cars in the relative comfort of the garage; aboard motorboats at the nearby Oder River; at the fire station because the ladder trucks were too long to fit into our curved driveway. A few customers had American automobiles which were —at that time—far superior to our German cars. Imported into Germany before the Hitler era, the Fords, Chevrolets, Buicks, Terraplanes, Essexes, Hudsons, La Salles and Packards were sturdier, more powerful and very much easier to work on than the great German makes—Mercedes and Horch, Opel, Audi and BMW. American cars also had modern conveniences which German cars did not have at that time: electric starters, hydraulic four-wheel brakes, directional indicators and self-moving windshield wipers.

Winter in Frankfurt/Oder was raw, snowy, cold and wet. To work in an unheated garage was plain miserable. We had an iron stove in one corner; the chimney and its bottom access were on the opposite end of the shop. The flue pipe sloped gently upward along the length of the garage wall from stove to chimney. Herr Schroth waited until our hands were blue and fingers frozen before I was ordered to "Thaw out the place!" The start of the risky maneuver to get a fire going began when I threw a match into a pail of oil and gasoline-soaked rags placed in the chimney's bottom, to create draft air flowing upward. The instantaneous explosion rattled windows, shook the garage—and me, too! When the stove was lit we not only heated the garage but we could place on top of the stove our garden watering can, in which we washed our hands. Did I mind this primitive life compared to hot-water faucets and radiators in our home, taken care of by the janitor? Not at all. This was a fire that *I* built!

A unique lesson I learned was that horse manure stops water leaks in the coolant system of an automobile radiator. On occasion I could be seen pulling a four-wheel kiddie-cart, armed with broom and shovel, tracking routes of horse-drawn milk, beer or ice delivery wagons as they made their slow daily rounds, bringing home a fine collection of horse apples. (Why does horse manure stop leaks? To this day I don't know. But such lack of theoretical knowledge didn't stop me in 1948 from the practical application of good, old-fashioned horse manure to the huge leaking coolant system of America's first sophisticated high-altitude, full-scale jet engine compressor test facility, then worth $6 million, at General Electric's engine plant in Lynn, Massachusetts. And with excellent results, I might add.)

When I was sixteen, I acquired my first motorcycle (a lucky

break since only one out of fifty-two Germans owned a motor vehicle in 1933)—a British Triumph, an indirect result of one of Hitler's mandates "to buy German products only." To discourage imports, the Reichs-Chancellor prohibited members of the armed forces from parking foreign-made vehicles on government property. Although Germany herself made excellent cycles—BMW's, NSU's and other makes—the British also made outstanding motorcycles—Nortons and Triumphs, for example. An army master sergeant, owner of a Triumph, was forced to sell his since he was not allowed to keep it parked at the barracks. Returning from a customer demonstration, he was killed in an accident: At a turn of the road, he was squashed between two streetcars going in opposite directions. The accident happened at night in front of our garage. Police cleared the street by shoving the crumpled iron pretzel into our driveway. Herr Schroth said to me next morning, "Boy, here is a pile of junk. You can get yourself a good motorcycle for nothing. If you want to work on it in the evenings, I'll give you the key to the shop." After three months of bending and painting, with substantial help from my master, I had a well-running machine of my own. It took less than five minutes of Schroth's time on the adjoining parking lot to show me how to ride it.

Like most young motorcycle riders, I became careless and nearly met the same fate as the sergeant, in a similar situation. Riding too fast downhill and turning closely behind a streetcar traveling in my direction, I, too, did not see the oncoming streetcar. Its bumper struck the rear wheel of my Triumph as I crossed in front of it. The impact threw me to the other side of the street. Miraculously, I got by without a fractured skull, or broken arms or legs; however, I suffered a deep three-inch gash above my ankle that left the bone bare. The painless wound took more than three months to heal, with lots of iodine and much care by our cook, who washed and wrapped fresh bandages nightly, keeping my injury secret from my parents. Of course, such a wound should have been stitched by a doctor. It left me with a half-inch-wide scar.

The only relief from physical labor during my six-day workweek was each Monday's eight-hour day at the trade school (four hours practical, four hours theoretical education). There I met young men from backgrounds and home environments entirely different from mine. We had two things in common: We were mechanic apprentices, and we were all wondering about girls, and what one was to do if one of them ever accepted a daring invitation to a date. . . . Naturally, we apprentices compared the masters for whom we worked. The general standard of teaching was high in every shop. The ethics of Alfred Schroth, his rules and standards, however, were the highest of any of the masters in Frankfurt, without doubt. I felt pride and satisfaction working for a man who really was a master of his craft.

When, after passing my journeyman's examination, I said good-bye to Herr Schroth, he said for the first time in three years, "Thank you, Neumann." It was also the only time he ever called me by name (first names were never used in prewar Germany, unless one was related or a very close friend). He shook my hand firmly, smiled a bit, wished me good luck and sent a bouquet of flowers to my startled mother. I felt sincerely grateful when I, in turn, thanked Herr Schroth—the man whom I had always addressed as Meister and who had given me a solid groundwork for what I hoped would be a rewarding engineering future.

3

Neumann Doesn't Fancy Feathers

In August 1914—the first month of World War I—all German males between the ages of eighteen and thirty were drafted. My father was thirty-three years old but he volunteered for military service, received his basic training and spent three years as an enlisted man in the trenches of France around Verdun and Cambrai. My birth in October 1917 was presumed to be the result of his 1916 Christmas furlough.

Following the armistice in November 1918, Papa returned to his business. The Norddeutsche Bettfedernfabrik (North German Feather Products), established in 1860, had become one of Germany's two biggest suppliers of feathers and down, plucked out of fowl on East European farms from Hungary to Finland. The feathers were cleaned in giant washing machines, dried and then separated by size in interconnected glass chambers with vertical baffles through which was blown a gentle breeze. The lightest feathers floated into the farthest compartments; the heavier ones fell nearer the entrance. Immense vacuum pumps sucked specified mixtures of down and feathers into huge sacks, for shipment to comforter and pillow manufacturers. Wet feathers stank to high heaven during the washing process, so badly that I was certain of one thing: It was no feather merchant for me!

During my twenty-one years in Germany, I saw little of my dad, whose life revolved around his business, except during a few nice summer evenings when he and I walked to the nearby infantry barracks and watched the changing of the guard. Both my father, Siegfried, and mother, Frieda, were born in Prussia. Both were of Jewish birth but nonpracticing; they considered themselves "Jewish Germans" rather than "German Jews." A Christmas tree, for example, was traditional in the Neumann home. My mother was an amazing woman for her time: skiing, ice skating, playing tennis, chess and bridge . . . and running the household with two maids. One member of her family was Dr. Fritz Haber, inventor of a process for making nitrogen, essential to Germany's ammunition industry during World War I. In appreciation, the Kaiser awarded him in 1916 the highest German civilian honor. In 1918, Dr. Haber won the prestigious Nobel Prize for Chemistry. My mother's only brother, an army captain decorated with the Iron Cross and commander of a mortar battalion, was killed in France six days before the war ended in November 1918. I had two elder sisters, Anneliese and Ulla, with whom I had as little to do as possible. Ulla developed cancer following a fall with her riding horse when she was only seventeen years old. The cancer was not recognized early enough and spread. She died at eighteen in 1932. The older sister, Annelie, married a dental surgeon in Frankfurt the same year and left for Jerusalem before Hitler came to power.

My hometown, Frankfurt/Oder, is the smaller of Germany's two Frankfurts. The very much bigger and better-known one is Frankfurt/Main, in West Germany. My Frankfurt, which straddles the Oder River fifty miles east of Berlin, is the seat of the government of the province of Brandenburg in Prussia. Its history dates back to the tenth century. It had its own university, was beleaguered by King Gustavus II Adolphus of Sweden in the seventeenth century, and was defended against the Russians by Frederick the Great of Prussia a hundred years later. The poet Heinrich von Kleist was born there, and the explorer Alexander von Humboldt studied at Frankfurt's university. After World War I, a good part of the 100,000-man Reichswehr (Germany's defense force permitted by the 1919 Treaty of Versailles to maintain internal order with limited equipment) was stationed in Frankfurt. The sleepy city on the Oder came alive each Sunday morning when a military band led troops to church services. Especially exciting was the cavalry band playing kettle drums and bugles; lead horses marched in step, providing a splendid spectacle every time they rode through town.

The Neumann home was a modern villa, with a big garden full of fruit trees and berry bushes. Discipline reigned there as it did in each Prussian household. You did exactly as you were told by your parents. There was no such thing as saying no to them! To whistle inside the house was absolutely taboo. ("Where do you think you are, in a stable?") You were not to have a hand in your pocket while talking to grown-ups. Some offenses invoked a ban on going to movies Saturday afternoon; others resulted in having my stamp collection locked up for a week. One time I shot a pane out of a street gas lantern in front of our house with my air gun; away went the rifle for thirty days. Showing any emotion in Prussia was considered sissyish. There was no kissing between parents and children—only a peck on the cheek before going upstairs to bed punctually at nine o'clock; and there was absolutely no crying. On the liberal side, I could do with my free time whatever I wanted, but *only* after my homework was completed. *"Erst die Arbeit—dann's Vergnügen"* ("First the work—then the pleasure") was an inflexible must. (I tried to introduce this healthy rule in my own home here in America thirty years later but lost the battle.)

In 1927, Charles Lindbergh succeeded in his historic trans-Atlantic crossing from west to east, exciting the imagination of the world. Although I wished that it had been a German who was the first to fly across the Atlantic, I was enthralled by Lindbergh's success in the single-engine *Spirit of St. Louis.* During that same year, everyone was fascinated by the attempts of several Europeans, particularly the French, who were trying to cross the Atlantic from east to west and seemed to be unaware of the strong prevailing headwind they had to fly into. None succeeded that year and several pilots were lost. (However, the following year, Köhl, von Hünefeld and Major Fitzmaurice—two Germans and an Irish officer—succeeded in the first flight from east to west, in a single-engine Junkers all-metallic aircraft.)

I was not quite ten years old in 1927 when a four-seater high-wing plane visited for three days a narrow grass strip outside Frankfurt—an overgrown runway of the 1914–18 war—offering fifteen-minute sightseeing flights for five marks. Without asking my parents for permission, I bicycled ten kilometers to the airfield and paid with money fished out of my piggy bank. From an altitude of less than 3,000 feet and at a speed of 55 miles per hour, I could easily locate our house and garden, trace the way I walked to school, recognize the library, sports stadium and ancient city hall, watch streetcars and pick out little dots I knew were people. Fear never entered my mind. During the bicycle ride back from the airfield I decided not to tell my parents. But at the supper table I couldn't hide any longer my adventure of that afternoon: I described how the town, our house and the way to school looked from the air. . . . The outcome was not punishment, but my parents' decision to take to the air themselves in a scheduled flight from Berlin to Dresden three months later!

In the same year I began to study for six years in a boys' Gymnasium (girls went to a Lyceum, whose surrounding streets were off limits to any schoolboy). Three foreign languages were mandatory: Latin for six years, French for four, and English or Greek for three years. Fortunately, I chose English. A mediocre student in languages and history, I was "very good" only in mathematics, physics, athletics and geography. Once I forgot the name of the river in India that flows from the hills of Assam, entering the Bay of Bengal near Calcutta. As penalty for not knowing its name, white-bearded Professor Dr. Neumann (no relationship) gave me a choice of either weeding the flower bed at his home one whole afternoon (a usual penalty) or writing three hundred times—neatly, numbered and in ink—the sentence "The Brahmaputra is in India." I chose the latter; it certainly was effective: I never forgot the name of this river and where it is. Ten years later, in 1942, I sent a postcard via the International Red Cross to my old geography teacher in Frankfurt/Oder's Gymnasium, telling him at that moment I was sitting on the bank of India's Brahmaputra wearing an American Army uniform. (The card was returned six months later: "Undeliverable.")

A painful adventure at the Gymnasium occurred when classmates persuaded me—as one who had had a clean record until then—to put half a pint of carbide from my bicycle lantern into the teacher's inkpot during recess. (In the late twenties, electric flashlight batteries just began to find their place in the market. Until then, a bicycle lantern consisted of a small bowl filled with carbide powder, onto which dripped some water. This produced a gas which the bike rider had to light with a match.) After class came to order, nothing special happened for the first few minutes. Then ink began to bubble over the pot rim. The old teacher, standing with his arms behind him and his back to the table, felt something moist crawling up inside his sleeves. It was a great joke for everyone but me; it resulted in my inability to sit down for hours. Our elderly teachers in their shiny, black threadbare suits ruled with an iron hand and a bamboo cane. But with the pain there came a certain pride in being "man enough to take it."

When the "official" school thermometer registered an outside temperature of 25 degrees C. (77 degrees F.) by 10 A.M. sharp, the school bell would ring three times and school was out for the rest of the day. Generation after generation of students at the hundred-year-old Gymnasium passed on a system to encourage the thermometer to climb another one or two degrees if there were indications that the forces of nature weren't cooperating. A lit candle tied to a broomstick was held below the thermometer, mounted high outside the north side of school, and a well-coordinated effort to divert the janitor's attention during the critical moment of heating the thermometer usually did the trick.

When I was twelve, one of my hobbies was the large family radio in the living room, with its wooden loudspeaker shaped like the one in "His Master's Voice" trademark. Not many people can recall how complicated radios were in the 1920s, powered by big batteries. Our set had a 100-foot-long antenna outside the house that had to be "grounded" by a hand switch during thunderstorms. It took some doing to tune in the station one wanted to hear, twisting five dials and plugging in different coils of wire atop the radio. The reward was the thrill of hearing musical station-signals from cities which seemed far away in those years—Rome, Paris, London, Moscow—even if one didn't understand a single word. As "radio expert" of the family, I wired an earphone extension into my bedroom. And when I was twelve, I built an automatic radio on/off device connected to an alarm clock; this saved me going down the unheated staircase at night to shut off the set.

Two radio events of the thirties stand out in my memory. I was glued to our set—as 125 million other Europeans were to theirs—when the annual Grand Prix automobile races held in six countries were broadcast over the whole continent in several languages. Each race had its own character: street races in tiny Monte Carlo; the thousand-mile cross-country race at Brescia in Italy; the Nürburg Ring race in Germany with its twenty hairpin curves and mile-long "straights." At every race it was a battle between Mercedes, Alfa Romeo, Bugatti and Maserati. I remember listening to the Nürburg Ring race in 1932, when German nationalism had begun to blossom. A white Mercedes trailed by a few hundred feet a red Italian car for each of the twenty rounds. Out of sight of the radio reporter, the German driver, Rudolf Caracciola, passed the leading Italian in the very last hairpin curve and came into sight ahead of the pack of competitors barreling down the last straightaway. The German spectators were as jubilant when the black/yellow-checkered flag went down as were the Italians when their soccer team won the world championship in 1982: the first German winner, in a German car, before a German crowd of half a million onlookers!

The other radio event took place in 1936. Between three and four o'clock in the morning, European time, the Max Schmeling–Joe Louis heavyweight boxing match was beamed from New York to Germany—an attempt that succeeded only partially. Radio transmission faded in and out as it traveled across the Atlantic. Everyone in Germany stayed up that night, suffering through the frustrating broadcast plagued by static. There was wild cheering (and no school that day!) when "our" world boxing champion, Schmeling, knocked out the American challenger in the twelfth round!

One Sunday in the summer of 1931, I watched a man unpack three bags and assemble their contents into a sleek kayak at a nearby lake. The boat was made of a lightweight frame of wooden ribs and spars put together in a particular sequence. The wooden skeleton fit snugly into a rubber-and-fabric skin requiring not a single tool. It was an 18-foot-long, two-seater Klepper, famous the world over. (A retired German Navy captain even used it for a solo crossing of the North Atlantic—in sixty-one days!) The folded kayak could be carried on streetcars and trains. Its design was truly ingenious.

I had saved enough money to buy such a boat when I was thirteen. (To buy anything "on credit" was unknown in Germany at that time.) My father agreed, subject to two conditions. Not only did I have to be able to swim but I also had to pass the German national lifeguard's test. I had tried to learn to swim for three years without success—primarily because I didn't like the cold, clammy harness attached to a leash which Frau Weingärtner, the ancient swim instructor (who had represented Germany in the first modern Olympic Games in Athens in 1896!), put around my chest. Whenever she was fed up with my not following her commands precisely on the "one-two-three" breaststroke rhythm, she dunked me—head and all—into the dirty Oder River. The lifeguard exam, in addition to requiring the demonstration of rescue and revival procedures, included jumping, fully clothed, from a 33-foot tower. This test was designed for grown-ups. My father was certain he'd be safe with me not getting the Klepper for a few more years, if ever.

From this challenge I learned a lesson far more important than how to swim: Give a person a desirable target and he will struggle like hell to achieve it, and he may succeed in doing the seemingly impossible! With visions of owning a Klepper as incentive, I learned to swim in one single lesson. I passed the national lifeguard test a couple of months later; at thirteen, I was the youngest person in Frankfurt to do so. Whereupon my father lived up to his word: He let me buy the Klepper.

For six years I paddled or sailed the kayak every Sunday, from spring to autumn. The two-seater sport boat gave me much self-confidence, independence and a pair of broad shoulders. No longer did I have to join other kids at the annual summer camp on a North Sea island, Norderney. Instead, I took to the water with a twelve-year-old neighbor—the four weeks of our July summer vacation. We had a small tent, charts, kerosene cooker, sleeping bags and lots of optimism! We kayaked on the Oder River to the Baltic Sea; pleaded with well-to-do Berliners to tow us behind their motor yachts; tented on shores of rivers and lakes of neighboring provinces; met poor and rich people, nice ones and not-so-nice ones. I learned to judge "go" or "no go" in many different situations, more about human relations in the real world—the give-and-take—than I could ever have studied in any book.

In the winter of 1932 I visited our city library, where I found books about powerful British aircraft piston engines—Rolls-Royce, Napier, Hawker Siddeley and Bristol. These were particularly interesting, since under the Versailles Treaty of 1919, Germany was banned from developing any aircraft engine above 150 horsepower. Why the librarian ever ordered such books (in English) I'll never know because I couldn't imagine anyone in Frankfurt caring anything about British aircraft engines. The loan cards showed that no one before me had ever taken them out. For me this literature was a gold mine: I began to learn about pistons and crankshafts, valves and carburetors, liquid- and air-cooled engines. Studying cross-sectional drawings and diagrams in these technical books was a big help to my future career.

At fifteen, I participated in a volunteer school program. The director permitted the basement of the Gymnasium to be converted into a workshop area where we could learn enough carpentry from a professional to build, to a proven set of drawings, a simple but safe full-sized glider. The two wings were transported separately to the flying field outside town and attached there to the fuselage. The pupil devoting the most time to building this "Grunau Baby" plane would have the first crack at flying it; then the next, and so on down the roster. My carpentry participation earned me the second flight position. With minimal theoretical instruction from the local glider club, each lucky schoolboy carpenter was strapped into the cockpit, to be launched 300 feet high into the air by a giant slingshot. Forty classmates pulled with all their strength the rubber rope laid out in a V; when there was sufficient tension and a bit of headwind, the flight instructor gave the signal for the brakeman to release the latch that anchored the plane's tail to the ground. The Baby flew! It was my first solo flight and surprisingly easy (and natural) to keep the glider in the air for two or three long minutes, followed by a not-too-smooth landing. . . .

Many people today know about the Hitler legend and his increasingly demented leadership which made him believe that it was divine intervention which made him, an ex-Austrian corporal in the German Army of World War I, the savior of Germany. In fact, however, he carried Germany toward her total destruction in World War II. In contrast was the favorable impact on the German people and foreign visitors of the glamorous, well-orchestrated and popular 1936 Berlin Olympic Games. Germany's #3 man, Hermann Göring—a very successful fighter ace in World War I and Commander of the reborn Luftwaffe—was put in charge of the Games by Hitler. When asked why three Jewish Germans were among those selected to represent Nazi Germany (one of them, Helene Mayer, was a good-looking blonde, daughter of a doctor, who had won a gold medal for Germany in women's fencing during the Olympics in 1928 and a bronze medal in 1932), he replied, "Ich *sage wer ein Jude ist*" ("*I* say who's a Jew"). Again, in the 1936 Olympics, Miss Mayer was a winner for Germany, earning a silver medal as an individual and a gold one as a team member.

The treatment of Jews was different in various sections of Germany. Up to the time I left Frankfurt for college in Mittweida in the mid-thirties, my school friends and apprentice colleagues did not hesitate for a moment to visit the Neumann home and hang their black or brown Nazi uniform coats on the clothing rack in our entrance hall. Our maid's boyfriend, an SS official, visited her in our home frequently. I talked at great length with black-uniformed SS officers (Hitler's elite corps) who brought their black official Mercedes touring cars for maintenance to Ehrenstein's (one of the three large Jewish-owned and -operated garages in Berlin where I worked during my college semester vacations). Just as other customers, the Blackshirts preferred this garage where their cars were well taken care of. Reports of major anti-Semitic activities began to circulate in Germany after the 1936 Olympic Games in Berlin had ended and the foreign participants and visitors had left Germany. Atrocities were committed by Nazis in some parts of the Reich on all kinds of Germany's population if they did not hew to the Nazi line—Catholic, artistic, scientific, Jewish—but not in other parts of the country. Things became much worse after November 7, 1938, when a young Polish Jew shot to death a German embassy official in Paris.

During those years, when what Herman Wouk has called the winds of war were blowing over the European continent, in Frankfurt/Oder the annual match between the local boxing club and the Jewish boxing club Maccabbi from Berlin went on as if nothing was happening. The "standing room only" crowds applauded and booed whoever deserved it and there were not visible any political overtones or racial bias.

4

Learning to Design for the Real World

In 1935, six months before graduating as a journeyman, I applied to enter Ingenieurschule Mittweida, Germany's oldest technical college. I was surprised and elated when word arrived from Mittweida: "Accepted."

Mittweida's buildings and laboratories were state-owned; nevertheless, tuition fees paid by the students' parents were high. The faculty consisted only of senior men; just as we students did, they came up via the "dirty fingernail" route. We recognized that those who instructed us knew what they were teaching or they wouldn't have been there. The Mittweida course material was different from that of the typical German *Technische Hochschule,* which, akin to MIT or Caltech, existed primarily to train research engineers in advanced problem-solving. Schooling of "practical" engineers, necessary to design and develop reliable, maintainable products for heavy-duty use, civilian or military, was the province of colleges such as Mittweida. The students' experience in handling hardware and in seeing failed parts firsthand, acquired during the mandatory apprenticeship, saved an enormous amount of time at the Ingenieurschule. Mittweida developed hard-boiled engineers who were trained to be skeptical of theoretical and overprecise calculations. The students were urged to design products as simple and accessible as possible; they learned to apply realistic design margins as a function of the particular product's expected use and to consider Murphy's Law: "Whatever can go wrong—will!"

At Mittweida, in addition to being taught standards of mechanical engineering, physics and heat transfer, we were trained to look at engineering drawings and instinctively to point out deliberately inserted errors—or quickly discover an omission—without having to go through detailed calculations. We were taught to "get a feel" for drawings laid before us. The question consistently posed was, Would it really do the job if it were built just as shown on this drawing?

The venerable and famous Ingenieurschule Mittweida had graduated many of the world's leading engineers, some of whom ran the Mercedes-Benz, Heinkel Aircraft or Zeppelin Airship works. Even the father of General Electric's former chief executive, Fred J. Borch, was a Mittweida student in the early 1900s. Further proof of the long-range success of apprenticeship/Mittweida-type education is the list of German-trained engineers who became business and technical leaders in America, Europe or even Russia after they emigrated to those countries following World War II—often not voluntarily but as part of the war booty. The first Russian Ivchenko jet engines, French ATAR fighter engines, and America's first axial-flow jets are examples of aircraft power plants that trace their heritage directly to Germany.

Mittweida had only 650 students, 15 percent of them foreigners. I believe it was a deliberate decision by the school directorate that foreign students at Mittweida were *not* required to have a practical mechanical apprenticeship prior to enrollment. Foreign Mittweida graduates would usually return to their homelands to serve in high government posts, displaying a German engineering degree only as a status symbol. Because they had never dirtied their hands or bloodied their knuckles in apprentice-type training before attending college, they did not benefit from the Mittweida-type education anywhere as much as did a German student.

The college had no campus living facilities and provided only a cheap lunch at the student cafeteria. My parents never once visited the school or the different boarding rooms where I lived. It was not lack of interest on their part; it was just not done by any parent. Telephoning home to tell the folks how one was getting along was unthinkable, an unwarranted extravagance. I wrote a letter once every two weeks (usually asking for some pocket money) and sent home dirty laundry if my landlady wanted too much for doing it.

At the close of each semester we routinely looked around for a "better but cheaper" boarding room with high hopes of finding a spinster landlady who would not only house us during the next term but also cook our main meal at noon and take care of laundry. For supper, which we made ourselves, we bought bread, jam and pickles; if one had lots of money, one could add cheese and sausage. I saved considerably by smearing jam on slices of bread, then scraping it back into the jar. That made the jam last twice as long, and enabled me to have cash for gasoline for my motorcycle.

Hardly any students bought new school books; the needed literature was obtained second-, third-, and fourth-hand from students who had graduated. One Saturday afternoon one of my housemates, who was going to leave Mittweida College prematurely, needed help to transport his heavy load of books for sale in the nearby big city of Chemnitz (today Karl-Marx-Stadt). This fellow, a friend and I dragged the suitcase weighing fifty pounds to the local train station and lifted it into the baggage net above the bench in one of the old-fashioned compartmentalized cars. The three of us were alone—until, at the last stop before Chemnitz, a tall, skinny farmer in a black suit with black top hat and his short, fat wife in her Sunday best climbed up the steep stairs into our train compartment. The man sat directly below the suitcase. A few minutes prior to the final stop, we began to pull the suitcase forward to the edge of the baggage net but lost control when the train jerked over some rough station switches. The heavy load slid over the edge and plummeted down smack on top of the farmer's hat, compressed it badly and rammed it over the man's head to below his nose tip, covering eyes and ears. Then the suitcase slid onto the seat. The farmer continued to sit straight as an arrow, silent while his wife began to scream for help. I nearly died laughing as two of us tried to wiggle the stovepipe up and over this poor man's face, while the third student held the man down by his shoulders. . . . When the farmer finally could peer from below the rim of his hat, he hit the first person he saw (who happened to be innocent) standing in front of him so hard that the young fellow fell against the door. I had heard before about people "dying laughing," but never thought that such indeed was possible.

There were two other Jewish engineering students enrolled in

Mittweida, one and two semesters behind me. Each of our fathers had been a volunteer frontline soldier in the German Army during World War I. Because the Hitler regime recognized such combat service, our fathers had provided us with "passports" which made us eligible for a college education. I personally was never insulted or aggravated by any Nazi apprentices or students. During my years at Mittweida, I experienced respect and friendliness from every other student, was invited to card parties, to join motorcycle rides into the country and to participate in kayak outings on Mittweida's lake. What did I feel about being friendly with these youthful Nazis? Nothing special. I knew that each of my German apprentice and college friends *had* to be a member of one or another of the numerous Nazi organizations—whether he wanted to or not. My grades at Mittweida were consistently *Ausgezeichnet* (Excellent)—often better than I thought I deserved, clearly indicating that the professors certainly were not prejudiced.

During 1935 to 1938, there were definite indications that Germany was going to launch a war as soon as she was "ready." The time would probably be sooner than estimated by Allied military intelligence, which was in the habit, I found out later, of underestimating enemy countries and their strength in men, matériel and will to fight. Hitler's aim was to become the dominant power in the world. The first stop was Europe. A popular German march song said it: "Today Germany belongs to us; tomorrow, the whole world!" When, in 1935, military draft became law in Germany, all engineering and medical students were deferred from immediate induction into any of the three services, but not from physical examination and classification. That classification was entered into the *Wehrpass* (military passport) every German over the age of twenty had to carry on his person wherever he went. My "class of 1917" was called up the first day of August 1937. We stood around the whole day for physical examination, without a stitch of clothing on, in a local restaurant closed to the public for the day. In Germany there were no exceptions or deferments: Everyone *had* to serve his country. If someone proved to have a fast heartbeat, an aching back or poor eyesight, there were plenty of jobs he could fill in the administration of the military and thus relieve a more healthy soldier for frontline duty.

However, if one of Germany's allies or customers needed her

technical skill and assistance—especially if it was a country which purchased a substantial amount of goods and generated badly needed foreign currency—the request for help was usually granted. This in spite of an acute shortage of the engineers needed for Hitler's massive rearmament program, to equip the military forces with superior artillery, tanks, planes, battleships, submarines and gasoline extracted and catalyzed from coal. The huge arsenal of modern weapons had to be developed practically from scratch and to be completed in six to seven years to meet the Führer's overall timetable, which contemplated the "provocation of an international incident" by 1939–40. This schedule was established as early as May 1935. In addition to the traditional arsenal of war, far advanced weaponry was a part of the overall get-ready plan: The development of sophisticated Tiger tanks, jet fighters, snorkel submarines, Stuka dive bombers, wooden troop gliders, V-1 buzz bombs and V-2 rockets was already far along. Even a German nuclear bomb was in an accelerated state of development. Hitler's aim was to create the most modern and powerful army, navy and air force the world had ever seen. It was *guns for butter,* and cost be damned. (The true German defense budget was declared a state secret; published military budget figures were falsified deliberately to fool the Allies.)

One sequence of events during my Mittweida years is seared in my memory: I saw the "big lie" technique, in Technicolor. After his troops marched in early 1938 into Austria, where the overwhelming majority of Austrians sincerely welcomed their German neighbors, Hitler's attention turned to Czechoslovakia. The Czechs, however, had military mutual assistance pacts with France, Britain and Russia. When reports about a possible German invasion of Czechoslovakia began to leak to the international press, Hitler recognized that he would have to negate those rumors quickly. He "wanted a peaceful settlement of the political differences" with Prague, he proclaimed over the German national radio—beamed to foreign nations. Hitler repeated this several times during hurriedly called meetings with Britain's prime minister Neville Chamberlain, France's Edouard Daladier and Italy's Benito Mussolini. He talked peace apparently so sincerely that just about everyone believed him; Chamberlain, returning from Munich to London, told the assembled world press at London's airport, "I believe it is peace in our time." He waved an agreement signed by Hitler and himself that very morning, promising "never again to fight a war with each other's country." (Hitler had kept silent about his detailed conditions for an acceptable "settlement" with the Czech government—a settlement that insisted on an immediate compliance with 100 percent of his demands.) To save peace at all cost, the Allies had agreed to the transfer of a large chunk of Czech territory to Germany, over the violent objections of the Czechs, whom the Allies had not consulted beforehand. While the "negotiations" about this "peaceful settlement" were going on, I looked from the second-story window of my room facing Mittweida's tree-shaded marketplace and saw below me masses of camouflaged tanks and artillery, plus troops with twigs stuck into the nets covering their helmets. Men and equipment were parked undercover, waiting in every available space in and around this little town in Saxony, thirty-five miles from the Czech border. The military juggernaut was ready to move the instant orders came from Berlin.

Only a few days after meeting with British and French leaders Adolf Hitler gave the go-ahead to move into Czechoslovakia and to break up that country whose government, firmly believing that Britain, France and Russia would live up to their commitments and march into Germany, had dared to refuse Hitler's ultimatum. Those countries did absolutely nothing but pressure the Czechs to accept Hitler's conditions—when it was already too late. German tanks crushed the border barricades; troops crossed into helpless Czechoslovakia. Thus, without a shot having been fired, Germany first had annexed Austria, then occupied most of Czechoslovakia a few months later, incorporating both its territory and its population into the Reich. Taking over this part of Europe went smoothly and in accordance with Hitler's timetable spelled out in a top-secret memorandum dated May 1938, which also called for the "manufacture of an incident along the German/Polish border that would justify massive military retaliation," a blitzkrieg (lightning war) against Poland to take place by August–September 1939.

A few weeks before the end of 1938, there was a typewritten notice on the college bulletin board stating that the Chinese Nationalist government of Generalissimo Chiang Kai-shek was looking for German mechanical engineers. Those interested were directed to contact the Chinese Embassy in Berlin. It also mentioned that arrangements had already been made by the Nationalist Chinese with Germany authorizing military service deferment for those who were hired. The notice was vague about the jobs, which apparently were somewhere in the interior of the Chinese mainland and were connected with defense. A job in faraway China sounded incredibly exciting. . . .

5

To Hong Kong by Air France—in Eight Days!

Between Christmas 1938 and New Year's Day 1939, I visited the Embassy of China in Berlin, where I saw the first Chinese in my life and was given an outline of what the ad on Mittweida's bulletin board was all about. China, for many decades on good political and business terms with any regime in Germany, continued to receive a great deal of modern German military equipment—even though Hitler (together with Mussolini) had signed the military Axis Pact with China's archenemy, Hirohito of Japan, only two years earlier. Military advisors also were sent to China, headed by the very able General Alexander von Falkenhausen. The country with the largest population in the world had been fighting an undeclared war against the Japanese invaders following the "Marco Polo Bridge" incident at Shanghai in 1937, provoked by the Japanese to justify their march deep into China. Generalissimo Chiang Kai-shek now needed engineers to teach his soldiers to maintain these new German weapons. The Berlin government was eager to continue its business with China, not only for old friendship's sake, but also to receive the foreign exchange generated by sales of military equipment. Besides, Nazi Germany operated an airline and a university in Free China. The Chinese Embassy staff were certain that there would be no problem for them to obtain a military deferment for those accepting the proffered jobs.

Terms of employment were attractive: a good salary paid half in German marks and half in U.S. dollars, and travel by air to Hong Kong with all expenses paid. The conventional boat trip from Genoa would have taken longer than seven weeks; an air journey with Air France was scheduled to take "only" eight days. The embassy would also arrange for all necessary transit visas through each of the sixteen countries in which the plane would stop en route to the Orient. Once in Hong Kong at the end of the flight, I was to report for further assignment to the Chinese Southwest Transportation Company (which, I guessed correctly, was a cover for the Chinese military in Hong Kong).

With excitement and anticipation I filled out the pile of visa application forms and sat for two hours in a narrow photo booth to take forty-eight passport pictures of myself. (Except for a trip to the Free City of Danzig in 1933, crossing the Polish corridor in a sealed train between East Prussia and my hometown, I had never before been out of Germany.) The mere thought of flying half around the world, at a time when flying for longer than a few hours was still most unusual, kept me awake long after going to bed. Although my parents did not believe that the trip would actually materialize, my mother made me read Pearl Buck's book *The Good Earth,* and *Six Hundred Million Customers* by Edgar Snow, both translated into German.

It was difficult to control my impatience and to wait one month for the travel passport and two more months until I obtained the last of the transit visas. My father engaged a tutor who spoke "real" English, having been once to London for a few months. His pronunciation of English turned out to be quite different from that I had learned years ago at the Gymnasium from teachers who had never left Frankfurt. In fact, at the first lesson, I couldn't understand a word that the tutor said—partially because of the moustache he had grown while in England. At long last, in April 1939, I was ready to go to the Berlin office of the world's most famous travel agency, Great Britain's Thomas Cook, with my passport and transit visas for sixteen countries: Belgium, England, France, French Tunisia, Italian Tripolitania, Egypt, French Lebanon, French Syria, Iraq, Iran, British India, British Burma, Siam, French Indo-China,

British Hong Kong and China! Cook's reserved a seat aboard a new French three-engine plane, the Dewoitine 338, that had been designed to Air France's specifications for its route to the Orient. (Only nine of these Dewoitines were built before the war stopped their production.) The flight was scheduled to leave Croydon Airport in London each Wednesday at 10:00 A.M. and to arrive in Hong Kong's Kai Tak eight days later—thus making good a daily distance of 1,200 to 1,500 miles. This broke up each day's flight into three segments because of the plane's range limitation. No other specific departure or arrival times except "day of the week" were scheduled for the whole trip.

The moment of departure from Bahnhof Friedrichstrasse in Berlin for Brussels and London came in May 1939, and with it a rare parental kiss and the needless advice *"Mach's gut!"* ("Do it well!"). It was the last time I saw my father, who died of a blood clot in 1940 in a Berlin hospital following a minor operation.

At that time nothing was known to the average German citizen of Hitler's secret strategy to sign in August 1939 a world-shaking ten-year nonaggression pact with his hated Bolsheviks (who overnight would change into good fellows). He had decided to launch a war immediately thereafter against Poland, with whom he also had signed a ten-year nonaggression treaty one year earlier, in 1938. Nor did we have any inkling that Hitler was intent on defeating the western European nations in a blitzkrieg after Poland's surrender and after having covered his back with the German-Russian treaty! Furthermore, he intended to tear up his nonaggression pact with Russia as soon as he had replenished the military supplies used on Germany's western front, then launch a war against the Soviets by mid-1941, thereby disrupting communication between China and Germany.

The Air France flight to Hong Kong could be boarded either at Le Bourget Airport in Paris or at Croydon in London. Years before at the Gymnasium, we read much about The City. I decided to take advantage of the opportunity and see for myself if Buckingham Palace, Piccadilly Circus, Nelson's Monument, #10 Downing Street and the London Bridge really looked the way they had been pictured in our textbooks. Surprise! I found everything was just the way we had learned in our English lessons including that indeed there were soapbox orators in Hyde Park, red double-decker buses, and very fast and deep underground "tubes." One exception: We had learned that Englishwomen looked like "stringbeans"; I found them quite attractive although slender compared to the plump German Fraüleins of the Hitler era.

The four days in London went by quickly. Air France's travel brochure for its route to the Orient, printed in French and English, highlighted points of interest one would see on the longest scheduled flight in 1939. It also listed technical details of this most modern French aircraft: ". . . the Dewoitine 338 is a 3-engined, all-metallic land mono-plane; has 12 Pullman-type seats which can be converted into lounge chairs; a cruising speed of 150 mph; a maximum altitude of 15,000 feet; its retractable under-carriage is delicately folded—in order to increase speed. The pilot's cockpit is like the control room of a vessel. Separated from the pilot's cockpit, the cabin is a calm and peaceful island where life, after farewells made at the airport, takes on the rhythm of a sea passage." The plane's crew consisted of pilot, copilot, navigator/engineer and a steward whose only job aboard was to see that each passenger fastened his safety belt properly and to serve refreshing drinks. All meals were to be taken on the ground; each night was to be spent in the best local hotel. Every morning, passengers with their handbags were to be weighed on a scale with a three-foot-wide dial (like jockeys with their saddles) and assigned seats by the flight engineer, to achieve a weight balance for the plane. We were admonished not to get out of our seats while the Dewoitine was airborne.

Marseilles was our first night's stop. There our Dewoitine was refueled from a gas truck for the last time. Thereafter, two men pumped aviation gas by hand out of fifty-gallon barrels rolled in front of the plane. A barefooted assistant, standing at first on the right, then the left wing, directed the gas hose into a large funnel covered by a chamois cloth to keep water and sand out of the fuel tanks. Passengers lunched, then rested in deck chairs (carried in the baggage compartment) in the shade under the wing while the plane was being refueled, which took longer than an hour. Twice on the trip, the flight crew themselves had to perform maintenance work on one of the two-bladed propellers which was causing problems. At the end of each day's flight in a different country, twelve touring cars pulled up at the plane's small exit door, a few steps above the ground. (It was taken for granted that each passenger had to have his own cab.) Native drivers would race from the airport to downtown, their hands on the claxon horn buttons and their feet trying to push the accelerators through the floorboards. People, goats, dogs, cows and camels crowding the streets before us would scatter in all directions. A feeling of total helplessness overcame me every afternoon when I was chauffeured by one of those suicidal cabbies, who all drove as if they had graduated from the same driving school.

We usually landed early enough to go sightseeing in town before dinner. Passengers and crew ate together at one long table in the hotel dining room. At the end of each dinner, the captain—first in French and then in English—told us when to be ready for breakfast (usually at 5 A.M., so that the plane's takeoff could occur in cool air, giving the wings more lift). It was after the second dinner that we noted our French captain and the only female passenger, a chic young French lady in slacks, leave the table and disappear arm in arm. They had developed, quite visibly, an affinity for each other. Or did the captain of our Dewoitine only carry out what the Air France brochure advertised? It said: ". . . later, conversation with fellow travellers, starting with observations on the sky, of the earth . . . the sensation of warm sympathy among people who fly together. . . ." This warm sympathy ended the morning before we reached Saigon, where a gentleman—rumored to be her husband and supposed to be the *gouverneur* of southern Indo-China—awaited the young lady with open arms.

The early morning takeoffs induced us to go to our bedrooms right after dinner. We turned on the switch of the two-bladed, low-speed ceiling fan common to all hotels in the Mid and Far East, gave the fan a shove with a bamboo pole standing in the corner of our rooms, to make it begin to rotate, took a cold shower and crawled into bed under the mosquito net. I learned by experience that unless one tucked the mosquito net firmly under the mattress from inside the bed, those miserable little bugs found their way through any tiny gap between mattress and net. A flashlight was always provided to assist in the nightly battle between the weary traveler and the few buzzing insects caught in the beam of light inside the net. The early-to-bed routine was not bad since not only were we awakened before sunrise but our clocks had to be advanced nightly by one hour as the plane headed eastward.

Countries which we overflew prohibited taking photos from the air. Our captain gathered all cameras in Paris and kept them in his custody. Likewise, our passports were collected and not returned until we reached our destination. Baggage inspection did not exist; not only because our suitcases remained locked in the plane's belly throughout the trip but also because hijacking a plane was unheard of in those days.

The route led across the Mediterranean Sea to Tunis, our first overnight stop in North Africa. I was startled to see men in flowing robes jostling each other in the crowded streets. The few Tunisian women we saw were covered from head to toe in garments resembling bedsheets, except for narrow slits at eye level through which the women could peer at the world outside. Camels, hundreds of donkeys and mule-drawn carts with two huge wooden wheels thronged the road from the airport to the magnificent Hotel Majestic. In total contrast was Benghazi, our next overnight stop, a seaport in Italy's Tripolitania (today western Libya). It was an orderly city with palm-tree-lined streets and a small modern hotel. At the Grande Albergo Berenici, built along the Mediterranean, I saw four German SS officers in black uniforms, drinking beer on the terrace. Those gentlemen were there, I assume, for an exploratory survey of the country through which General Erwin Rommel's Afrika Korps was to head on its way to Egypt for its rendezvous with British General Bernard Montgomery three years later.

After refueling at Alexandria, the next stop, we flew on to Beirut (the capital of French-administered Lebanon), where we were driven downtown for lunch at the beautiful Hotel St. George. We enjoyed the spectacular view from the most easterly point of the Mediterranean. Even more spectacular were the shapely Frenchwomen (in bathing suits briefer than any I had ever seen) who lounged under sun umbrellas while enjoying their drinks.

After takeoff at Beirut, we circled for half an hour trying to gain enough altitude to cross the mountain range to the east. The plane was unable to clear its ridges. The Dewoitine was forced to return to the Beirut airport; two hours later, when the air was cooler and supported the wings better, we made it, but barely. It was a short hop to Damascus. Spending one Arabian night in Baghdad sounded exciting—yet I was disappointed not to find one sexy belly dancer; Iraqis liked their women fat! Once we landed at Basra, in southern

Iraq, whose runway was nothing but a totally unmarked two-mile-long stretch of sandy beach, 300 feet parallel to the Persian Gulf. Here folding lunch tables were set up in the shade of one of the plane's wide wings; four barefooted waiters served spicy dishes while two young boys tried to keep hundreds of flies away by waving feathered fans in front of our faces and over our dishes! It took nearly two hours to pump the gas out of the many fifty-gallon barrels into the Dewoitine's tanks. With no air traffic whatsoever in this area of the world, air controllers were not necessary. Our plane simply taxied to the end of the beach, turned around, revved up its three engines and took off—just like that. At Karachi, India, the next stop, we disembarked in front of the world's biggest hangar, built nine years before to shelter the British R-101 dirigible, larger than even the German Zeppelin *Hindenburg.* But the R-101 never made it to India: On its maiden voyage in 1930, from London to Karachi with British Secretary of State for Air Lord Thomson aboard, the overweight airship ran into very strong headwinds and had not sufficient engine horsepower to make good forward speed. It struck the ground near Beauvais in France, killing everyone aboard. (Four years later, I would be working in that very same hangar in India on a secret job as a Technical Sergeant in the United States Army Air Corps.)

My first stay in India, though brief, left me with indelible impressions of people in throngs even larger than they were in the Near East; masses of men and women were crowding the narrow streets, most of them barefoot and dressed in rags. Hordes of beggars, old and young, were pleading for "baksheesh, sahib!" About half of the women I saw in India were not hidden under bedsheets as they were in North Africa or Iran. They balanced baskets of fruit, bags of rice or even blocks of cement on their heads, walking absolutely erect, their bodies draped in simple saris. These Hindu women had a red dot in the middle of their forehead, indicating, I was told, that they had fulfilled their religious duties that day, and many of them wore a small golden ring through one nostril. Just about every young to middle-aged woman had a baby on her hip. At night many homeless Indians slept on sidewalks or in doorways. Some lay body to body on the cool tile floors of the small Indian air terminals, so that we had to step carefully over the sleeping forms in the dimly lit buildings on our way to the weighing scale and our Dewoitine before six o'clock in the morning. I had never before heard about "sacred cows," but here they were in person, so to say. White cattle walked or lay unmolested in the middle of the roads, slowing down traffic and stopping streetcars until conductors succeeded in coaxing the animals off the tracks. No Indian struck a white bull or cow. The same scenes were repeated whether in Hyderabad, Allahabad or even Calcutta. A walk through Calcutta, which had the most elegant shopping district in India (replete with glass showroom windows and tea salons), often required the pedestrian to make an arc around the cows stretching themselves out comfortably across the width of the sidewalk. I also saw members of the Calcutta sanitation department at work: Huge, ugly black vultures dropped from the sky anywhere in town to pick up garbage, scraps of food or dead animals! Without inhibition, men washed themselves under fire hydrants or relieved themselves in the open street gutters. Standing at the banks of the wide Ganges or Brahmaputra rivers, I couldn't believe what I saw: Bloated human bodies floated downstream, giant black vultures getting a comfortable sail on the corpses and a free meal as well. Not all dead bodies wound up this way; others were burned downtown on six- to eight-foot-high wooden pyres, at any time during day or evening.

We spent the night at the Great Eastern Hotel in Calcutta, which, like the Raffles in Singapore, was a world-famous landmark. The difference between the quiet, cool and gracious living inside these hotels and the miserable condition of the poor and undernourished masses outside was shocking.

It was not as bad in Bangkok, where we stayed at the Oriental. In contrast to the elaborate, gilded, colorful temples across the Chao Phraya River, Bangkok houses looked like shacks. Siamese people were poor but not as bad off as Indians. The incredible humidity in Siam's midsummer, the curried food, people hanging like grapes on the outside of the few slow-moving streetcars, and the change of races, customs and languages at about every plane stop a few hours apart made me feel absolutely unable to absorb what I was seeing. There was so much going on in this part of the world, so many sights for which I was totally unprepared. How many more nonwhites there were than whites! Still, I felt apart from all this: I was a mere traveler en route to China. Crowds and misery were left behind when our captain advanced the three throttles of the French engines and our Dewoitine climbed into the pure, cool and empty dark-blue sky. . . .

Before leaving Bangkok, the captain handed out our cameras and announced we would be flying low over neighboring Cambodia and would see en route one of the seven wonders of the world, the recently discovered ruins of Angkor Wat. He gave us a bit of history. Hundreds of years ago a whole civilization lived and worshiped in the temple city and then somehow disappeared. Dense jungle grew so completely over the abandoned city that it was not visible any longer. Now, he told us, archaeologists had discovered and were restoring Angkor Wat. We circled over the site at less than 1,000 feet altitude so that each of us could take photos from the air. I took only one shot, then was out of film. Dammit! While other passengers kept clicking away, I rewound and reloaded my camera as quickly as I could, yet it was a lengthy process with the old type 120 film. After circling several times over blackened stone walls and towers, a moat around the temple, a long bridge crossing it and an occasional working elephant moving trees, our captain pulled aside the curtain separating cockpit from cabin, to be sure that everyone was ready to proceed with the flight. Noticing my camera problem, he called reassuringly, "Take your time, we'll wait!" and kept on circling to the right and to the left until I, too, had my fill of photos.

We landed at Saigon, a beautiful city which had to my mind the world's most exotic women, a blend of French and Chinese charm —until they opened their mouths and revealed their blackened, betelnut-stained teeth! Sweet-scented shade trees lined the avenues. Mild, fragrant evening air and crowded Parisian-style sidewalk cafés, tiny-waisted women wearing slinky silk gowns, young girls in colorful shirts hanging loosely over black pantaloons, and their always neat hairdos—all this made Saigon seem a paradise. Conical bamboo hats, worn by farmers in the rice fields against sun and rain, were also worn as decoration on the backs of the shapely girls in town. Streets were busy with pedicabs whose drivers sat behind their fares. No cows, camels or buffaloes blocked traffic in downtown Saigon. I well understand now what made French Foreign Legionnaires so eager and delighted to be stationed in Indo-China rather than in France's north or central African colonies.

It seemed many months and not just a few days ago that I had left Europe. Except for Saigon, all African and Asian cities at which we had stopped were overcrowded, but all seemed reasonably well run and orderly in spite of the masses of pushing humanity. Much credit must go to the British, the French and the Italians for governing their colonies, for providing communication, police and court systems, bank and postal services, and for training native people to take over more and more of their own countries' administration. If they had only done something about birth control!

On the eighth day after leaving London, right on schedule, we landed on the grass runway at Hong Kong's Kai Tak Airport located on the Kowloon peninsula (the other half of Hong Kong is the island on which is located the city of Victoria). Kai Tak was international in character. German-Chinese Eurasia flew their three-motor Junkers planes twice a week into Hong Kong from that part of China's interior not occupied by the Japanese; Pan American-owned Chinese National Aviation Corporation (CNAC) operated its twin-engined Douglas DC-2's and DC-3's, flying from Hong Kong to the interior of China and on to Burma on alternate days. British Overseas Airways' Sunderland flying boats from London landed in Hong Kong harbor every Monday. Air France brought their Dewoitine to Hong Kong every Thursday. Twice a week, Pan American Airways' famed Sikorsky Clipper flying boats connecting San Francisco with Manila continued on to Hong Kong.

All but three of the twelve passengers who had boarded my Dewoitine in London and Paris had left the Air France plane in Indo-China. Now three taxis took each of us to Kowloon's magnificent Peninsula Hotel, which was as typically colonial British as could be. This was the last prepaid night. The following morning, June 1, 1939, life in the Orient began for me in earnest.

I inquired at the desk of the elegant lobby where I could find the Chinese Southwest Transportation Company responsible for my continuing trip into the interior of China. The concierge was told by the city's phone operator that the line to C.S.T.C. had been disconnected ten days earlier. No forwarding address was left behind. Perplexed and more than a bit concerned, I changed my few remaining English pounds into Hong Kong dollars, wrote an airmail letter to the Chinese Embassy in Berlin explaining my predicament (of which they were probably unaware) and requesting further instructions by cable. As directed by the military before I left Berlin, I reported my arrival and my whereabouts to the German Embassy within twenty-four hours. They promised to check into the situation with the Chinese government. Estimated time for my letter to travel to Germany by airmail was about three weeks. Since I had been issued in Berlin only a transit visa through Hong Kong, valid for six days, I needed, first of all, permission to remain in Hong Kong. Next priority was to find a job until the odd situation with the Chinese cover agency was cleared up.

It was a lucky coincidence for me that Far East Motors/Far East Aviation Sales and Service was located directly across the street from a side of the Peninsula Hotel. FEM/FEA had responsibility for the assembly and repair of Chevrolet cars and trucks for Hong Kong and all of South China; FEM/FEA was also an agent for British Standard automobiles. In addition, they sold and serviced light biplanes at Kai Tak Airport. The general manager was Mr. Claude White, who had run away from his California home when he was fifteen and was hired aboard a three-master to sail to China in 1900. He was the first American I ever met. A tall, sporty, good-looking, white-haired man in his early fifties, White wore neither tie nor jacket. After hearing my story, White said he saw no difficulties in getting me a work permit from the Hong Kong government valid until my departure for China whenever that would be—provided I worked in his garage for an admittedly very low HK $30 a month (U.S. $7.50!). His miserly salary offer aside, I liked White instantly; he was not pompous like the few Britishers I had met. "Okay," I said to White's surprise, "I'll take the job." What I liked most about him was that he acted promptly: He immediately telephoned a few people while I waited in his simple office, then asked me when I could start work. "After lunch," I replied. White offered to have someone find a place for me to sleep for the next few weeks, and added that he would take care of the rent until I knew what was what with my job in China. (I found the "Let's do it right now" attitude to be typically American—something I liked and admired about the few Americans I was to meet in the following eighteen months.) To be realistic: I had no choice but to accept his offer since I was totally out of funds. White gave me the opportunity to survive until I could locate the Chinese Southwest Transportation Company. (He told me after the war, which he spent in Japanese internment, that he wanted to see if I was seriously interested in working with my hands!)

By nightfall I had a furnished room, hot as hell and without a ceiling fan, but it had a bed and was within walking distance of Far East Motors. White loaned me a few HK dollars to buy some cheap clothes, much lighter in weight than those I had brought with me from Germany. Within a few days I became accustomed to the Chinese mechanics at Far East Motors, and they to me. They taught me to swear in Chinese and I taught them to swear in German. Although White had a shop manager from Macao (a nearby Portuguese colony whose main industry was gambling) who spoke English, Portuguese and Chinese fluently, White asked me a week after I had started to work at FEM to select from an American catalog tools and equipment which I thought his place really needed. Soon Far East Motors had a more professional look. Lines were drawn where cars were to be parked; tools were now stowed in orderly fashion when not in use. I introduced gas and arc welding into Far East Motors, built a frame for a chain hoist on rollers, to remove and install engines. (Until then, this had been done by coolies with bamboo poles over their shoulders!) Word got around amongst European (mostly British) car owners that there was a "white man working with his hands"—which was incredible for those who lived in the Far East before World War II. New customers showed up and wanted their cars repaired by "that European"; some of the cars to be serviced even arrived at Kowloon from Victoria on Hong Kong Island via the Yaumati car ferry. How lucky that my father had arranged for me to learn the trade at Herr Schroth's in Germany!

When I received the first monthly paycheck in my life, I returned it to the front office because there was obviously an error: HK $90 instead of the HK$30 I had been offered. White's only comment: "No mistake!" At the end of the following month (it was then July 1939) I was paid HK $270! Enough money to repay White what I owed him, to move into a more livable place with a ceiling fan, a few blocks farther down the road, and to afford to quit eating green bananas for breakfast, lunch and dinner! (Those "green" bananas were ripe, sweet and typical for South China.) One day I developed a bad toothache and was going to visit a Chinese dentist who had his office around the corner of the garage and who was known to pull teeth for a small fee rather than to treat them. White heard about this. "You have an appointment with that American dentist across the street in the Peninsula Hotel at eleven o'clock." I objected that I could not afford such a dentist; I was told by White in no uncertain terms to get out of his office; he had already agreed with the American dentist that he, White, would take care of my bill. What a man!!

When he knew me a bit better and my English had become more fluent, Mr. White invited me one Sunday morning to his beach cottage in the nearby New Territories (a strip of land surrounding Kowloon, forming a buffer between British Hong Kong and Japanese-occupied China). I was to meet his Viennese wife and to go swimming in the South China Sea. A few weeks later, he asked me to a small party he was giving at his home. White wanted me to tell his guests about my extraordinary flight from London to the Orient. This forced me to struggle through my travelogue in English, a matter made somewhat easier because the Americans were drinking continuously! One of the lady guests asked if I had also begun to speak Chinese. To prove that I did, I let go with a melodious phrase I had picked up from my mechanics, who used it at just about every happy or unhappy occasion. After hearing me blurt out the Chinese words in their proper intonation in front of all his guests, White pulled me aside and asked if I knew what I had said. "No," I replied. White advised me strongly "Never again" to use *that* Chinese phrase in the presence of ladies.

Hong Kong was like heaven for me—and to most white people who lived there. I particularly enjoyed watching the lively harbor crowded with watercraft of every description: hundreds of junks with huge brown or purple canvas or bamboo sails as far as the eye could see; motorized wallah-wallahs (water taxis) crossing the harbor; freighters from all over the world; British warships and submarines moored to buoys in the middle of the harbor or tied to docks; warships of many other nations taking on coal, water or fuel; gleaming super-modern white German tourist boats and the old-fashioned four-funneled yellow American President liners; new colorful French and Dutch ships with single rectangular funnels that exhausted diesel fumes rather than coal smoke. The six most indispensable craft in Hong Kong, the famed STAR ferries, plowed back and forth across the harbor in 1939 (and still do so in 1984), one leaving punctually every five minutes from the famed Clock Tower terminal on Kowloon and another one from the pier at Victoria's business district. STAR ferries had two classes. First class was on the upper deck, immaculate, used primarily by whites and the more affluent Chinese: its fare was ten HK cents. Second class, on the lower deck, was jammed with coolies, chickens in bamboo baskets, bales of cargo, and those who—like myself—appreciated its three HK cent fare. (In 1939, four HK cents were worth one U.S. cent.)

Hong Kong was protected by 50,000 British Empire troops stationed on Kowloon. Every so often, a Scottish band marched down Nathan Road, the main street of Kowloon, in kilts, playing bagpipe music. To me they neither looked nor sounded very military, accustomed as I was to German uniform pants rather than skirts, and Prussian fife, bugle and drum corps bands. English, Scottish and Indian Army personnel lived in barracks along Nathan Road or in their own apartments. Even the lowest-ranking British Tommy had his family living the life of Riley in this low-cost city, where an amah to care for the babies and a houseboy to wash the car could be hired for a mere pittance.

Every single night the evening sky over Hong Kong provided a breathtaking experience: When the sun had just settled behind the mountains and rocky, uninhabited islands, the blue sky grew paler, then changed into a rainbow of every color imaginable. This spectacle, combined with the panorama of huge Chinese characters ablaze in colorful neon lights along the length of the harbor advertising beer and watches, fountain pens and insurance, sparkling in the reflecting waters of the China Sea, made it clear why Hong Kong was called the Jewel of the Orient. In 1939, in all of Hong Kong there was only one high-rise building: the twenty-story Hong Kong & Shanghai Bank on Victoria. Two large magnificently sculptured lions guarded its entrance on either side. Many foreign companies had offices in that gleaming white brick-and-marble building, which one year later was to play so significant a role in my career.

Just when my life began to take on an orderly pattern, war in Europe became a distinct possibility. Large black headlines in Hong Kong's *South China Morning Post* announced the surprise signing of the German-Soviet ten-year Nonaggression Pact, startling everyone of every nation and preparing the world for the beginning of World War II. One week later, on August 31, 1939, Hitler's regime faked a Polish attack on the Silesian radio station at Gleiwitz, by dressing twelve inmates of a concentration camp in Polish uniforms, shooting them and distributing their bodies around the station. For four minutes a faked Polish radio address had been delivered from that German station. He used this "incident" as the "final reason" for the massive German invasion of Poland on September 1, 1939. England and France had done nothing but issue complaints and warnings when Germany occupied first Austria, then Czechoslovakia in 1938—in spite of their specific mutual assistance agreements with Prague. Thus Hitler was encouraged to go on with his plans to smash Poland (with which England and France also had a recently reconfirmed mutual aid pact). Against the advice of his military staff, Hitler took a calculated risk that the Allies ("democratic worms who are afraid to fight") would once again do no more than threaten him. He directed the Wehrmacht to use all its available power, including that normally stationed at Germany's western front, so that Poland would be "wiped out" quickly before London and Paris could respond effectively. On September 1, 1939, Hitler received an ultimatum from England to recall his troops, which had crossed the Polish border at 4 A.M. that morning, by noon of September 3, 1939, or "England would find herself at war with Germany." France made the same demand. I doubt if anyone in England or France really expected the Germans to pull back their rapidly advancing and victorious troops.

At 7 P.M. Hong Kong time, on September 3, 1939, the ultimatum expired and England *was* at war with Germany! A few minutes later a British immigration official, accompanied by a British officer and two beturbaned Indian soldiers with rifles, knocked at my door and demanded politely to see both my travel and military passports. I was driven in a bus to police headquarters in Kowloon, where twenty-five other Germans had already been rounded up. More were brought in by the hour. At 2 A.M. there were ninety of us, all men. The British apologized profusely for the inconvenience they caused, and I believe that they meant every word they said. We were driven in three buses to LaSalle College in Kowloon. Under floodlights, Chinese workmen were driving stakes into the ground to surround the college with barbed wire. Four watchtowers were being built also. Its dormitory was empty since there was a school vacation. A few more German businessmen were brought in during the early morning.

It became obvious that the British had not given any thought to the possibility of interning one hundred Germans. Theoretically, we were really not prisoners of war but rather civilian internees. One Englishman told us that we would leave this temporary camp soon, but no one knew anything for certain. Most of the Germans had been residents of Hong Kong for many years, possibly longer than some of the British, and they spoke English fluently. The majority had not even been to Germany since Hitler had come to power. Incredible as it may seem, two old Chinese were also interned with us; in their younger years, they had been houseboys in the former German colony of Tsingtao in northern China and had there accepted German citizenship in 1906. They were unable to speak a single word of German!

One of the more touchy problems for the British was that of race: One white man had to show respect for another white man in front of the hundreds of "yellow" Chinese, who, curious, crowded outside the barbed-wire fence of our camp and peered inside. To prove that we were not "real" prisoners held by the British, we were permitted to hire Chinese kitchen personnel who prepared and served meals in the dining room, washed dishes and did our laundry. Life in the internment camp was not bad at all! We sunned ourselves, played tennis on the school courts and enjoyed watching the British guard climb down from one of the four watchtowers to retrieve a tennis ball someone had hit over the fence.

Our relations with the British camp commander were excellent. Pretty soon a discussion was held between our elected representative and the commander about setting up a bar in the college auditorium. These negotiations were successfully concluded. Well-to-do German merchants would pay for construction, furnishing and initial stocking of the bar, which was to be open for Germans every night from eight to nine and for our British guards from nine to ten. After only a few days of this arrangement, the Germans invited Tommies for an "early drink"; since British soldiers were paid very little, the Germans' invitations were gratefully accepted. Someone then had a bright idea: Why not combine the two one-hour sessions into a common "bar open" time from eight to ten for both guards and inmates? Why not? said the German representative. Why not? asked our British commander. Guards had orders to leave their heavy Enfield rifles outside the bar leaning against the wall of the corridor leading to the bar entrance. And so it went, peacefully, intelligently and thirst-quenchingly. Until . . . one of the Germans, slightly drunk, played a stupid prank: The man grabbed one of the rifles and shot out a light. Pandemonium broke loose; we were ordered back into our sleeping quarters. . . .

As a consequence, not only the commander but his whole guard force were replaced by a very much tougher Scottish commanding officer and his men. The bar was closed for good; no more playing tennis; no longer were we permitted to employ Chinese houseboys. New camp rules were established including the "calling of the roll" every morning before breakfast. It went like this: We 102 inmates formed three ranks; the Scottish officer, his sergeant major at his side, stood before us facing his enemy aliens. Each internee had been assigned a number; when his number was called by the sergeant, that internee stepped forward, stood at attention and called out his name. His number and name were then checked off on the sergeant's clipboard; the man could leave for breakfast. All went smoothly until my turn came. My number, which was assigned at random, was called; I stepped smartly forward, said "Neumann" and was ready to go downstairs for breakfast. But the officer said some words to me in his Scottish burr—none of which I understood. There was silence. Nothing happened for a few moments, then the officer yelled at me, pointing to the rear. I stepped back. Again my number was called, again I came forward and gave my name, but once more the lieutenant let loose a harangue. I simply could not figure it out and turned to my fellow internees, asking them if they knew what the problem was. Yes, they knew, and told me in German. "Step back, wait for your number to be called, step forward, give your name and be sure to say sir!" When my number was called for the third time, I stepped forward with confidence and in a ringing voice called out, "Sir Neumann!" The commander was furious, the Germans collapsed, laughing. (After all, we *had* learned at the Gymnasium in Frankfurt that "sir" means *Herr.*) I was dismissed and sent on to breakfast.

Weeks of dull camp life followed. It looked as if the European state of war would continue for several more months, even though

Poland had surrendered after fighting bravely and unassisted for less than three weeks. No real combat had erupted on the German western front during the first few months of the war; British and French military missed their chance to hit Germany while her troops were busy in Poland. The British in Hong Kong wanted to empty LaSalle College so that its students, back from summer vacation, could again use their quarters. A military court in camp interrogated each of us individually about our backgrounds, why we were in Hong Kong, et cetera. A few days later, the camp was closed and the college reopened to its students. We were not free, though: We were restricted to our places of work during the day and to our homes after 8 P.M. Hong Kong's beautiful harbor was declared off limits to us. Each semiprisoner was issued a pink ID card, with photo and fingerprints, marked "Enemy Alien" and had to report to the nearest police station every Friday afternoon. Furthermore, a British resident had to guarantee the alien's safe behavior. Mr. White was delighted to have me working for him again and had persuaded Sir Lawrence Kadoorie, a well-known British financier of the Sassoon Bank conglomerate and the head of China Light and Power, also one of my customers at the Far East Motors garage, to vouch for me.

I became White's chief mechanic; my predecessor, Ling, accepted gracefully my replacing him. Before internment I had workbenches installed at Far East Motors so that Chinese mechanics, who worked squatting on their haunches on the cement floor, could instead work standing at benches. When I returned from internment, I found no one sitting on the floor all right—but sitting on their haunches, *on* the benches. Thoughtful Mr. White had me elected a member of a modest British club next door to his garage so that I could eat one warm meal each noon. Far East Motors became a home for me. I was almost afraid that my frequent attempts to establish contact with the elusive Chinese Southwest Transportation Company would be successful.

One day White called me into his office, where a British officer was seated. The Governor's British Daimler, similar to a Rolls-Royce, and the only one in the colony, needed a major overhaul, especially of the preselecting automatic transmission, which jumped gears every so often. Since no one in Hong Kong had overhauled a Daimler, "how about that German trying his best?" Great! I did the job myself: I carefully disassembled every single part of the complicated automatic gearbox, sketched a step-by-step record on a long roll of tough British toilet paper, ordered replacement parts from London, which were flown in by BOAC (British Overseas Airways Corporation), and then reassembled the whole works. After engine and transmission functioned smoothly in the garage, the British major who had delivered the Daimler rode next to me, both of us sitting on a wooden orange crate wired to the car's chassis (I had not yet reattached the body). We drove for three enjoyable hours through town and into the hills of the New Territories surrounding Kowloon. The officer invited me to lunch and we two had a jolly good time! A few weeks later I was asked, this time by a high-ranking British officer, if I would care to overhaul his Napier aircraft engine installed in his racing motorboat tied up at a dock in Hong Kong harbor. He was going to get permission for me to do the job there.

I was looking forward to seeing once again my favorite harbor at the end of June 1940 when a British soldier walked into our shop and handed me a sealed letter. I opened it excitedly and read: "Compliments! His Excellency the Governor of Hong Kong requests you to return to LaSalle College immediately, and to leave the Crown Colony within 48 hours after midnight. Should you not do so, you will be transferred to a newly established central internment camp in Ceylon, India. Yours respectfully, Your Most Obedient Servant, His Excellency the Governor of Hong Kong." I could not quite believe what I read and thought that there must be a mistake somewhere. "Compliments!"? Did his Daimler run *that* well? But then the rest of the message did not make much sense. I took the letter to White's office. He made a few phone calls and then gave me the bad news: There was nothing he or anyone could do. All Germans had to leave Hong Kong.

The reason for our expulsion was simple: France had just collapsed under a sudden, overwhelming German military onslaught following on the heels of the quick surrender of Denmark, Holland and Belgium. The disastrous English withdrawal from Dunkirk with the loss of just about every major piece of British military equipment had to be blamed on somebody. The most plausible excuse for the Allies' debacle was the presence on Allied territory of German tourists called the "fifth column," suspected of sabotaging communication and transportation systems in those countries. We now had two days to get out of Hong Kong—but the British decided to keep our passports! Adding injury to insult, those of us who wanted to try to go elsewhere than Ceylon had to pay five Hong Kong dollars per day for an armed soldier who was under orders to follow his charge everywhere. I was one of the few who tried. I went to the American, Swiss, Philippine, Japanese and Portuguese consulates, and sought, for the last time, to locate the Chinese agency to which I was to have reported one year before. No dice—it had left without a trace. I also never heard again from the Chinese Embassy in Berlin, which probably never did receive my letter. One fact was clear: No country would admit anyone without a passport. I tried General Motors and Ford, which had assembly plants in the Orient; again, no success. The German-Chinese airline Eurasia had discontinued service into British Hong Kong the day the war started, and the German consular officials had departed for Macao a year ago, so there remained only the Pan American-controlled Chinese airline CNAC (Chinese National Aviation Corporation), which flew daily from Hong Kong to Free China's wartime capital, Chungking, then on to Kunming in southwest Yunnan. At four o'clock in the afternoon of my second (and last) day before shipment to Ceylon, my British guard and I went to the Hong Kong & Shanghai Bank building.

That is where I ran into W. Langhorne Bond in the elevator. Seersucker jacket over his shoulder and tie loosened, Bond made a few calls on my behalf. Just like White, he did so right then and there . . . no delays! I could hear only his side of the phone conversation. "No, that's too late." And after a few minutes of what seemed to me hours of silence, the American sitting before me said, "Okay. I'll have him at Kai Tak at eleven P.M. Thanks." What had happened? "I just hired you for CNAC. Normally I can hire only Americans, and these only to instruct Chinese. You'll be permitted into China without your passport. We'll fly you to Chungking and then on to Kunming, where you can make your Chinese contact, the one you missed here last year. The Chinese Air Force is directed by an American former officer, Captain Chennault. Look him up, because after you arrive at Kunming, CNAC will dismiss you—since our airline has no work for you. All I want is to help you out of here. The rest is up to you. Good-bye—and good luck!"

A miracle had happened; a coincidence in timing. I was now able to get out of Hong Kong without any documents! Haughtily, I told my guard, who had been banished to the corridor, "Tell your officers they don't have to worry about Neumann any longer. All they have to do is get me to Kai Tak Airport before eleven o'clock tonight."

A few minutes before midnight, I was airborne once again. The colorful bright lights of Hong Kong disappeared below me. We were over China! Just lucky, I guess. . . .

6

Dealing with Wheels in Yunnan

The sun came out just as we skimmed slowly over the country 1,000 feet above the ground. Farmers wearing coolie hats worked knee-deep in muddy rice fields; others were guiding ancient plows dragged at barely perceptible speed by huge black water buffaloes with five-foot-wide horns. Men were pedaling or hand-cranking wooden wheels to transfer brownish water from one rice paddy to another. Row upon row of women, also wearing conical bamboo hats, bent down in the muddy fields planting rice seeds. The scene must have been the same a thousand years ago.

I had considered my stay in Hong Kong as "living in China." But what little I saw, flying low over Chungking, China's wartime capital, and during the few hours I spent on the ground there, compared to what I had left behind in Hong Kong the evening before, was dramatic. A real shock were the contrasts between Chungking's mud houses, their roofs protected by layers of thick bamboo poles one foot apart (to detonate Japanese bombs before they penetrated the roofs), and Hong Kong's sturdy four-story colonial-type brick buildings; dilapidated rickshaws versus hundreds of spotless, feather-dusted automobiles and taxis; masses of people in Chinese garb versus residents of the British colony dressed in Western-style clothes; battered buses without fenders, without headlights and sometimes even without engine hoods versus Hong Kong's well-kept red double-decker buses and green double-decker trams; a jungle of muddy streets compared to orderly roads with white center lines. Would *all* of Free China look like what I've been seeing during the last few hours here in Chungking? I asked myself. What on earth am I doing here?

We had an early lunch at the Chungking airfield, on a narrow island in the brown Yangtse River, where our plane stopped for several hours. After one year in the Orient I finally had my first all-Chinese meal. (In Hong Kong I had eaten Western food only.) It was also my first serious bout with a pair of reluctant chopsticks. From now on I was to savor exclusively the pleasures of Yunnanese, Cantonese or Mandarin meals, all eaten in customary help-yourself style. Whatever one chose to eat, one course after another including the meal-closing rice and soup went into the same small bowl, one for each guest. A kind Chinese lady passenger showed me how properly to hold the two wiggly chopsticks made of bone; she taught me to lift pieces of chicken, a glassy-eyed duck's head, slices of pork and Chinese vegetables out of large bowls placed by the waiter in the center of a round dining-room table, typical for Chinese restaurants, into one's own personal bowl. She also showed me how to eat rice properly: Hold the bowl to your lower lip, then use your chopsticks to shovel rice into the open mouth. Efficient, if not elegant!

The cabin of our twin-engine Douglas DC-2 cooled off quickly after takeoff from humid, sticky Chungking. We climbed through a thick layer of brown dust that hung in the air 1,000 feet above the ground, the result of Japanese bombing of Chungking the night before. Our CNAC captain was an American ex-Navy flyer; his copilot, a Chinese. The DC-2 climbed to 12,000 feet and headed southwest for the province of Yunnan. After bumping along in heavy overcast for four hours, we broke through the clouds and flew low over a reddish-brown terraced landscape with some dry and some water-filled patches rimmed by narrow green footwalks. Rice paddies dug into the slopes of the hills were contoured to take advantage of every square foot of the "good earth." It was hard to realize that the peaceful-looking country below had been at war for over three years in a struggle against a superior enemy. Japanese troops had succeeded in isolating China for all practical purposes, except along her jungle frontier with French Indo-China and the mountainous border with British Burma.

Our final approach to Kunming was low over a smooth lake; the plane rolled to a stop on the grass runway. In a grove of pine trees along the airfield was the headquarters of the Chinese Air Force. It seemed months ago that Mr. Langhorne Bond had urged me in his Hong Kong office to visit with the chief advisor to that Air Force, retired American Army Captain Claire Lee Chennault, yet it had been only yesterday afternoon.

The instant I stepped off the DC-2, I liked the 6,000-foot-high Kunming. Cool, dry late-afternoon air and two rows of tall populars framed a road leading to somewhere—my first impressions. Wooden airplane maintenance shacks with signs in Chinese, English and German stood within a few hundred feet of our plane. A large tri-motor German Junkers JU-52 with EURASIA and a huge black swastika on a red/white background painted on both sides of its fuselage, a plane identical to those used by the Luftwaffe as troop carriers in the war in Europe, was parked nearby. Kunming was the second-largest city in unoccupied China; it was the capital of the province of Yunnan, in the deepest southwest corner of the mountainous country. Kunming had an estimated population of 150,000, consisting of longtime residents, refugees from the Japanese, and tribespeople who had moved down from the northern hills.

We passengers walked toward the bus leaving for town. An English diplomat who had sat behind me on the flight from Chungking explained that in Yunnan people had actually seen airplanes years before they ever saw their first automobile. No wonder that neither farmers nor buffaloes had looked up when we approached the grass runway!

Our rattly bus bounced along the five-mile road from the airfield through the center of Kunming to the French Hôtel du Lac, where all transient foreigners stayed. Everyone on the streets wore a Chinese outfit in gray or blue, or a green military uniform. Coolies (*coo-li*—man with power) pulled rickshaws over the cobblestoned streets, pushed two-wheeled carts or carried two baskets swinging at each end of a flexible bamboo pole on their shoulders. They wore shorts, straw sandals, conical "coolie hats" and rags around the neck to absorb perspiration. Walking in tandem, two coolies transported squealing pigs whose legs were tied together, hanging head down from bamboo poles. People strolled on sidewalks or in the middle of roads, scared aside only by the honking of a rare automobile moving at slow-motion speed. These cars belonged to government bigwigs, bank presidents, members of the eight consulates in Kunming, or to the few other foreigners living outside the town wall.

Except for two picturesque fifteen-story-high pagodas, none of the mud and wooden houses along the tree-lined streets was higher than two stories. Kunming's main street was lined with restaurants, one movie theater (showing ancient French films with Chinese subtitles) and stores selling locally woven cloths and straw sandals, vegetables and rice. All shops and restaurants were wide open to the main street. Nowhere existed a door or a pane of glass; small windows were covered with rice paper. An irresistible aroma of freshly baked French bread drifted out of bakeries and tea shops. Signs and advertising bore Chinese characters only.

Yunnan's official language was French, yet barely anyone could speak it. Also French were the poilu helmets of policemen dressed in black uniforms and gray tennis shoes, standing on cement pillboxes at major intersections to direct traffic and to shoo people and animals to the side of the roads when a lone auto or bus came along. The French influence originated primarily because of the Michelin, a narrow-gauge, single-track, winding and climbing railway clinging to the sides of the mountainous terrain between Indo-China's (now Viet Nam's) Hanoi and its Chinese terminal, Kunming. Connecting the two cities, this railway was built seventy-five years ago by daring French engineers from the sea level at the Gulf of Tonkin to the 6,000-foot-high Kunming. It now carried all military and civilian goods Free China had received since the British caved in to Japanese demands to stop all further assistance to Free China via Hong Kong. One of the casualties of British policy was the Chinese Southwest Transportation Company (my supposed employer), which was forced to disappear from the Hong Kong scene without a trace—only days before I arrived there in 1939.

At my first dinner at the Hôtel du Lac, struggling with chopsticks, I sat at a table next to a German diplomat and learned about the political and military situation in China as *he* saw it. Germany continued to support both Japan and Free China, the latter with advisors and armament shipped via Russia and the Trans-Siberian express; and the Japanese received the latest data and drawings of advanced weaponry via German and Japanese submarines commuting between Brest (the German submarine base in occupied France) and Yokohama. Despite the official Axis (Germany–Italy–Japan) Pact, Germany's heart, he said, was still very much pro Chiang Kai-shek. Had not one of Chiang's sons just graduated from Germany's "West Point" at Potsdam? And had not Japan been a member of the Allies in World War I, occupying the two German colonies in China, Tientsin and Tsingtao? The Deutsche Universität, teaching all its classes in the German language, continued to function in Free China though it had already been moved rearward twice because of Japanese advances. Besides, Herr Lutze, chief pilot of the airline Eurasia, had been shot up very badly in his clearly marked German transport plane by a Japanese fighter the past month. Obviously, Germany had good reason to help China.

The diplomat explained to me that the governors (or warlords) of the dozen or so provinces in Free China collected their own taxes, had their own military and police forces and made their own laws. It was Generalissimo Chiang Kai-shek's tough job to forge the many provincial units into one Nationalist army under the unified command of the central government, which was advised by German officers. According to the diplomat, Chinese people were not feeling any impact of the unimportant Communist movement under Mao Tse-tung in the north, since at that time both Nationalists and Communists were cooperating in trying to stop the Japanese invaders. He added that a Chinese puppet regime—under Japanese control—ruled "occupied China" and its 450 million people.

Early next morning, I took the bus back to the airfield to meet Chief Advisor Claire Lee Chennault. Chennault was related through his mother to Civil War general Robert E. Lee; he was a tough-looking, square-chinned man with jet-black hair and piercing eyes. This man was a former schoolteacher from Louisiana and later an enthusiastic pursuit pilot in the U.S. Air Corps. After his retirement, he had been called in 1937 to China by Mme. Chiang Kai-shek. His name had been given to her by Americans flying in China, and by a Chinese Air Force commander who had visited the U.S. Air Corps base where Chennault was developing a fighter tactic he deemed essential in any future war. Chennault's unconventional approach brought him into conflict with higher-ranking American officers who favored bombing missions and downgraded the fighter role.

When I visited with him, Chennault spoke little, looked at my German records and told me what he believed the Chinese government would face in the next few years: China would be isolated and cut off from all supplies. The Japanese would see to that. Therefore, a lifeline route, the Burma Road, was being built—and just about finished—by thousands of coolies armed with shovels, picks and sticks of dynamite. The road was a narrow unpaved highway, winding across incredible terrain, mountains and valleys, swaying bridges . . . and not a single service station along its five hundred miles to the Burmese-Chinese border village of Wanting. Chennault, who was given the rank of colonel in the Chinese Air Force, urged me to see Herr Nothmann, German manager of the local French Renault/Tessier truck assembly plant which was under contract with the Chinese Ministry of Transportation. I could do more for China at this point in time, he said, if I would help her ground transportation. The German equipment I was hired to maintain was "being buried in the hills" by Generalissimo Chiang Kai-shek's men, for the inevitable civil war to come, between Nationalist and Communist forces, once the conflict with Japan was won. Although Colonel Chennault offered me employment by the Chinese Air Force he felt it was too early for that; he could always get in touch with me later on. He then hand-wrote a note to Mr. Chang, president of the Central Bank of China in Kunming, asking him to help me lease one of the sixty-seven "houses for foreigners."

Chang assigned to me the last available unit of the European-style homes in the bank's new "Model Village," half a mile outside the city wall. CNAC pilot Bob Angle and his wife, an ex-U.S. Navy nurse, lived in House #66, and Colonel Chennault in House #65. I got #67! The other occupants of Model Village homes were all Europeans, doctors and engineers, working directly or indirectly for Chiang Kai-shek's government.

The Model Village was encircled by a high mud wall ensuring safety and privacy. Each of its small, modern bungalows was simply but adequately furnished. In addition to two bedrooms and one living room, each bungalow had a fireplace, a bathroom with Western-style toilet (but without a tub or shower), glass windows and even a glass door leading to the terrace with garden. Every morning water for the kitchen and bathroom had to be carried by coolies in buckets from the nearby Dianchi lake, up a narrow steep gangway to an elevated reservoir on stilts, one for each house. This water, of course, could not be drunk without being boiled for a few minutes. A female cook and a live-in amah, their long pigtails swinging at their backs, worked in my home a seven-day week. They were provided by the Bank of China, which had an agreement with the central government to take care of "friendly" foreigners. Rental fee for the bungalow and servants was not even discussed. Electricity was available only during daytime from 6 A.M. until 6:30 P.M., when the BBC (British Broadcast Corporation) radio news was over.

A few months after I moved into the Village, a U.S. consular official named Brown, who had received orders to return to the States, left me his large Dutch Philips shortwave radio in exchange for a brake repair I had performed on his powerful Hudson convertible. I, as did all foreigners, listened religiously to the 6 P.M. BBC broadcast beamed from London to the Orient. The excellent newscast began with the ringing of Big Ben's bells and their newscaster's famous deep voice: "This is London calling. Here is the news. . . ." It was our only source of reports from the outside world. Three cheers for London's BBC!

At the Renault plant outside of Kunming, twenty diesel-powered trucks were being readied for their first trip over the not-quite-finished Burma Road. "Would you be willing to lead a convoy of trucks to the Burmese border?" asked the manager after I had introduced myself. The trucks would be loaded with bars of tin and wolfram, to be shipped to the USA from China via Rangoon, capital of Burma. The convoy would return with barrels of aviation fuel and diesel oil, boxes of machine-gun ammunition and five-hundred-pound aerial bombs. A report of my findings on road conditions, with appropriate recommendations to the Ministry, would be appreciated. I was fascinated and agreed.

"When do we go?"

"In two weeks." My co-driver would be Chinese, as were all thirty-eight drivers of the nineteen other trucks.

I went daily to Renault to assist in the checking out of each truck and its engine. To my great disappointment, I found that the Cantonese language and my favorite, action-producing profanity acquired in Hong Kong were of no use whatsoever in Kunming. In many cases, a totally different pronounciation of identical Chinese characters made an understanding between myself and my coworkers impossible.

A week before the convoy's departure, the wailing of an air raid siren indicated to Kunming's citizens that "something was up." The Chinese, with Chennault's help, had developed a simple but superb warning system. One five-foot-diameter red ball was hoisted on a mast standing on the town's highest hill, visible in all directions, indicating that "unknown aircraft" were heard or seen two hundred miles away from Kunming, apparently heading for that city. A second ball was hoisted forty-five minutes later, when the "unknown aircraft" were heard or seen fifty miles away, apparently heading for Kunming. A third red ball, ten minutes later, was added to the other two already hoisted when the aircraft were spotted near the capital and presumed to be Japanese. No further movement was permitted on any road. Three V formations of nine silver dots shining in the sun, barely visible, were droning at this moment right above our heads outside the city wall—having passed over the center of Kunming seconds earlier, at 22,000 feet altitude. A glitter in the sky below the silver dots looked as if leaflets were fluttering to the ground. But then, a faint whistling sound grew to a crescendo! Air began to vibrate until three strings of explosions, racing along the ground at 200 miles per hour (the same speed as that of the planes), blasted and shook the earth. Those dots were no leaflets; they were bombs that had glittered in the sun! Each explosion looked like a small erupting volcano. It was my first "getting bombed" experience—and it was also the first mass bombing of Kunming. That day I was more fascinated than scared.

After thirty minutes, the first air raid was over, the "all clear" was sounded and the red balls lowered. Later I walked to town to see what had happened . . . and was horrified. The center of the "open city" (unprotected and of no military value) was strewn with dead people, their innards ripped open, displaying unbelievable colors; wounded men, women and children were lying everywhere, bleeding; crying people were kneeling next to their dead family members; wooden houses were burning. Two water buffaloes, torn to pieces by shrapnel, were scattered across the sidewalk near the North Gate, through which I had entered.

During each raid that followed, I was scared stiff, fearing selfishly that one of the Japanese bombardiers might release his load aimed at downtown Kunming seconds too late and thus smash to bits our Model Village, located half a mile away from the city wall. Miscalculated strength of head or tail wind between the planes and the ground could also have done us in. So could bombs hanging up a split second too long in the planes' bomb bays. Neither Chinese houses nor Model Village homes had basements. We foreigners hurried from our houses and huddled for protection in a four-foot-diameter cement drainpipe underneath the road leading past our village when the second red ball was hoisted. Chinese in our vicinity would press close to us foreigners: They were convinced, a farmer told me, that we "white devils" would know where the bombs would land, and where to take cover for safety.

Kunming was hit at precisely 10 A.M. nearly every day except Sunday. Twenty-seven Japanese twin-engine planes in three V formations of nine each, dropping five bombs per plane, used Kunming as a training target, according to the testimony of a Japanese pilot captured later. The Chinese population was amazingly resilient; by early next morning, they had disposed of the dead and cleaned up rubble that blocked the main streets. Men and women shoveled heaps of collapsed mud walls into big piles, mixed them with water and began to form new clay bricks, laying them out to dry in the sun for three days. If not hit again within the next seventy-two hours, those bricks would be used for the immediate rebuilding of their destroyed mud houses. Thanks to rice paper windows, there were no injuries from flying glass splinters.

The town had no water system. Kunming's fire department in action was like a scene from a Gilbert and Sullivan opera. Pails of water to extinguish the flames were handled by a bucket brigade involving more than five hundred men. Buckets were passed from hand to hand over the long distance from the Dianchi lake, then returned for refilling. The fire chief, decked out in a colorful uniform and a chromium-plated helmet, stood on an elevated platform under a large standard indicating to everyone from where orders were issued and to where reports were to be directed. The adaptability of Kunming's population showed also in other ways. After two more ten-o'clock raids it became obvious that daily bombings could become a way of life; without panic, people evacuated their town between eight and nine o'clock each morning, streamed in long lines through the four gates, herding their animals and carrying on bamboo poles whatever possessions could be packed in baskets or loaded into carts. A few miles beyond the city wall, entrepreneurs set up road stands with food which they cooked or baked on the spot. After the hand-cranked "all clear" siren sounded, people returned to town to see if they still had a home or a place of business.

Although Kunming had to adjust its life style, the foreigners' Model Village was barely affected. My cook and amah, and their colleagues in neighboring homes, did not even leave the house during the raid. Naïvely, they crouched under the kitchen table for protection, covering their heads with large kitchen pots. The maids did household chores until they heard the "urgent" air raid signal, and had dinner ready at five o'clock as if nothing had happened. A few unlucky police and firemen had to remain in the empty city because there were thieves on the first days of bombings. Stolen valuables had to be repurchased by the owner at the official thieves' market that same evening, even if he could prove original ownership. The logic was that the sellers themselves had to pay the middlemen (usually the thieves). When a thief was caught, justice and harsh punishment followed before nightfall: Leg irons, connected by a short chain, were riveted over the criminal's bare ankles, rubbing them raw and bloody; he had to carry in his hands the proverbial large iron ball attached by another heavy chain when he was led around town. Such exhibition was enough to deter most other would-be criminals.

In spite of the daily bomb attacks, life in Kunming had its humorous aspects. I recall the ingenious way in which waiters lured more generous tips (which they pooled) from their restaurant clients. The name of each departing customer and the amount of his tip were announced in a booming voice by the headwaiter to all diners before the guest was able to leave. Such a procedure sure got rid of lousy tippers!

At one time or another, everyone in town needed rickshaw service: two-wheeled vehicles pulled by lean and strong coolies. These simple, uneducated, illiterate but honest fellows had their own professional code of honor and pride. They never argued for a higher fare once a price was agreed upon beforehand, even if the ride turned out to take twice as long as estimated, because of a bombed-out bridge, for example. "A deal is a deal, and a handshake a commitment." Coolies would refuse to accept a tip. To pull heavy customers up steep hills was an achievement. Most foreigners were heavier and bigger than the Chinese for whom the rickshaws had been designed, especially massive Frau Heinrich, the wife of a German engineer working for the Chinese government. She was endowed with an oversized rear end to match the rest of her body and had literally to squeeze herself into the narrow rickshaw seat. Sometimes she had to bounce her posterior a few times until it settled down. On one of her trips to town, the rickshaw coolie failed to consider her high center of gravity and was unable to negotiate a sharp turn in the road while running full speed downhill. It took several minutes to extricate Frau Heinrich from the overturned vehicle lying on its side, with the help of bystanders including myself, who happened to bicycle past the place of the accident.

After three weeks of checking equipment and loading, my convoy of twenty cab-over-engine Renault trucks departed for the Burmese border. Fine-tuning of sensitive diesel fuel injection pumps had taken longer than the plant manager had estimated, because his Chinese mechanics lacked experience with diesels. It was obvious that if one of the overloaded vehicles would begin to slide on the road, incredibly slippery because of the mixture of brief daily showers with its red clay surface, one would not be able to stop the truck from slowly heading sideways over the cliff, dropping hundreds of feet into nowhere. Doors and windows, even the drivers' windshields, had therefore been removed before departure from Kunming, permitting a hasty exit should it become necessary to abandon the Renault before it disappeared over the road's edge. On dry days, too, there would exist constant danger for drivers and their trucks: Sandy stretches of the Burma Road easily crumbled under the concentrated weight and pressure of each wheel. Blocks of hard teakwood had been fitted between each truck's chassis and axles to permit driving with the heavy overload, preventing the trucks' springs from breaking.

It was estimated that it would take eight days to complete the 580 miles' distance to the Burmese frontier and eight days to return to Kunming, averaging 5 to 10 miles per hour, with frequent stops to heave out of one's path the boulders which had dropped from the steep sides of the mountains next to the road. The losses expected by the government were one truck out of a twenty-vehicle convoy per trip.

Chinese drivers were an ingenious and cheerful lot who invariably came up with practical solutions to "unsolvable" problems. For example: One of the five main bearings coated with soft metal failed inside one engine, forcing my whole convoy to stop for a day. I spent hours underneath the truck's motor with its oil pan removed, lying in the mud and wondering what to do. One of the drivers suggested cutting his felt hat into ribbons, then wrapping these strips around the crankshaft to serve as the bearing in lieu of the original soft bearing metal that had melted. True enough: The oil-absorbing felt did the trick! Nine days after we had left Kunming, we reached the border town, Wanting—having lost only one truck owing to a slide. Its crew were able to save themselves.

I had been told that an Australian adventurer was the convoy leader of twenty British trucks driving from the Burmese frontier east into China to Kunming. I worried in advance how our convoys could possibly pass each other, truck for truck, when we would meet and how long it would take for each of "my" trucks to pass the twenty going in the opposite direction! And how *did* we pass the convoy coming up from Burma? Each lead truck halted as soon as it caught sight of the other. We convoy leaders walked toward each other, looking for sections in the road 18 feet or wider, to permit two vehicles to inch past each other, so close that some truck bodies literally had to be pried apart. Since nobody volunteered to have his vehicle on the side with the steep drop when squeezing past another, we agreed that each truck stay alternately on the "inside" (safe) and then on the "outside" (dangerous) side of the road. On the trip back we did better: no vehicle loss at all. I returned to Kunming dirty and unshaven, but pleased about this unique experience.

The day after, I debriefed Renault's management and then the Ministry of Transportation. I suggested dynamiting "passing spots" into the mountains at regular intervals. I happened to be still in that government office when a shiny black 1940 Packard limousine drove up. The gentleman driver—a rarity in China's interior, where professional chauffeurs did all the driving—introduced himself to me as the consul general of French Indo-China. Would I be willing to examine and tune up his engine, which had not been running smoothly since the car had been shipped by train from Hanoi? Not only would I get paid "well" for my efforts, but he would recommend me to other members of the diplomatic corps.

The rough and dirty Renault truck yard was not suitable for work on this fancy car. I punched a hole into the mud wall next to my Model Village home, wide enough for the Packard to drive onto my tomato patch. Primitive, I admit, but there was a war going on.

Shortly after the Packard purred again the way a Packard should, many of the 150 cars registered in Kunming lined up before my house—their arrival accelerated by the certainty of daily air raids. The owners wanted to assure themselves of safe and reliable transportation, to get their families out of town before the 10 A.M. bombing. Pilot Bob Angle, my neighbor, seeing the cars parked in front of our two houses and down the road, proposed becoming my silent partner by taking orders for needed spare parts, then bringing them up from Rangoon on his twice-weekly CNAC flights. I accepted his offer and gave notice to Renault's Herr Nothmann. Angle and I decided to run a garage business until the Chinese Air Force called me. We rented an open space not far from my house, had a straw shed built to protect a dozen automobiles from rain and sun, and had a few pits dug into the earth to work more easily underneath the cars. Two weeks later, "Reliance Auto Service" was born and in full swing. I painted the automobile on the garage sign myself; a local friend drew the Chinese characters. Within three months I had as customers not only all members of the diplomatic corps, but also the International Red Cross, British Friends' Ambulance trucks, and Kunming's Chinese VIP's; and, of course, Colonel Claire Lee Chennault's 1936 Ford.

The standard sickness of Kunming's automobile engines was water blockage of minute passages in the carburetor or fuel filter, the result of water in gasoline barrels stored too long in backyards by the owner of each car. Such water was a consequence of condensation of the air layer between the surface of the fuel and the inside top of the barrel, due to wide temperature variations between the hot days and cool nights in Kunming; the less fuel left in a fifty-gallon gasoline barrel (thus creating space for more humid air), the worse became the condensation problem. A single drop of water in an automobile's carburetor can cause an engine to run rough or not at all!

The work was often difficult because of the lack of special tools and limited time to perform repairs between the daily raids. I hired two dozen good mechanics. They were as loyal and hardworking a bunch of men as I ever had work for and with me. I personally participated in repairing cars, from dawn to dusk. Every morning, promptly at nine, anticipating the ten-o'clock air raid, Reliance Auto Service drove and towed a string of cars from its garage, parked them underneath trees along the highway and kept on with repair work until bombs began to drop and we had to take cover in an adjoining road ditch.

My mechanics and I learned to understand each other very well: Language ceased to be a problem because they were primarily Cantonese, whose dialect I had learned to speak in Hong Kong. In addition to the VIP's who had become my customers, U.S. Navy and U.S. Marine convoys bringing supplies from Rangoon to the two American gunships lying at anchor in the Yangtze River near Chungking stopped regularly at Reliance Auto Service for check-ups. (Kunming boasted a red-light district, which delighted the American sailors; work on their convoys in my shop gave them an excuse for a two- to three-day layover to get their own engines tuned up.) Had I worked as a chief surgeon in any leading hospital of the world, I could not have earned greater respect and appreciation for my work than I did as grease monkey with dirty fingernails, in blue coveralls, in the province of Yunnan! When the first automobile had arrived there in 1938, it was a prestige item; by mid-1941, however, it had become a matter of life and death for owners who relied on their cars to get them safely through the town gates before the bombs came whistling down.

My Chinese mechanics performed repairs which any expert in the States would think were absolutely incredible. For example: A broken main leaf of the front spring of a 1937 Ford belonging to the British consul general, Prideaux-Brune, made further use of his car impossible. The long, thin steel spring had given up its ghost in one of the thousands of potholes in Kunming's roads. No replacement spring was available in Rangoon. It was the end for his car unless we *made* one of those flat, 5-foot-long, 3-inch-wide, ¼-inch-thick spring leaves, elastic and shock-absorbing, out of raw material available in Kunming. We could purchase only long, round steel bars. My blacksmith, Kung, had brought to my garage his tools of trade: anvil, sledgehammer, and an empty fifty-gallon barrel converted into a charcoal-fired furnace; its air bellow was made of a hollowed tree trunk with plunger, and a chicken's wing with its feathers as air valve. Kung was from North China, a skinny six-foot-tall fellow; he came with an assistant: his Cantonese wife. She was short, well rounded, with two large breasts to which she held a four-month old baby that she nursed continuously and unabashedly in the presence of everyone in the garage.

Kung heated sections of the round steel bar white-hot, then yelled at his wife to quit nursing the fat baby. She practically dropped it on the dirt floor against the mud wall, grabbed a fifteen-pound sledgehammer, and with incredible strength—in a "round swing" (prohibited in Germany because of its centrifugal-force danger!)—she slammed the heavy hammer over and over onto the hot iron rod held by her husband on the anvil. This process was repeated for two days. It was fascinating to watch Mr. and Mrs. Kung make a flat spring emerge out of the round bar.

Finally came the critical moment of tempering the spring leaf to make it hard yet elastic—but not so brittle that it would break again. "Impossible without heat-treating equipment," experts would say! So did I, but not skillful Mr. Kung. He had placed a long narrow trough in the corner of my garage; each of our employees had been asked, for the past several weeks, to relieve themselves into that container. It was half full when Kung dipped the white-hot spring leaf into the urine "just at the right moment," lifted it out of the resulting steam for a few moments, watched the steel's color darken and then lowered the leaf into the yellow liquid, first a little bit, then altogether. . . . I just stood there and shook my head in disbelief. Two years later, the Ford and its front spring were still doing well, and the British consul general was a very happy man!

One afternoon a rickshaw pulled up at my garage. A well-dressed gentleman of obviously high rank (one could easily identify a VIP by the louder bell on his rickshaw and its air-filled tires) asked me in good English—on behalf of the chief of police of Yunnan, Mr. Li Tse Cho—if I would, "in return for a princely sum," examine his brand-new 1940 Buick Century, which only a few days ago had arrived via the Burma Road from the docks at Rangoon. "Unfortunately, to the undescribable sorrow of my master, the new car is deadly sick, and he is afraid to ride in it; its engine makes a dreadful noise. Would you please have a look at this limousine and see if it could be saved?" I agreed to visit the Buick next morning, before the first red warning ball went up the air raid mast.

The VIP rickshaw was sent to pick me up early. Dressed in blue coveralls and holding on my lap a bamboo basket with tools, I was pulled to the chief's mansion near one of the two pagodas surrounded by a high wall. A wrought-iron moon gate swung open seconds before I arrived. My rickshaw was pulled into a paved driveway past a row of automobile stalls. In the last stall stood the "sick" Century limousine, the largest Buick. Just ahead of it was a streamlined, green four-seater Peugeot convertible, vintage 1939, with green leather interior. A beauty!

In front of his Chinese mansion, Chief Li and his chauffeur-officer were awaiting my arrival. It was a scene out of a storybook. After Chief Li and I exchanged bows, an aide began to translate for the chief. Li Tse Cho (*Li*—the good) thanked me very much for coming. The interpreter then gave the driver's version of the car's illness and its symptoms. It made a terrible noise the few times he had driven the chief and his family through town into the country, at the hoisting of the first air raid ball. The Buick was not used on any other occasion; for his town travel, Chief Li continued to use his rickshaw. I suggested a test drive, with the chauffeur at the wheel to demonstrate the sound the interpreter had tried to describe. Chief Li got into the rear; I sat next to the driver. After only a mile through city traffic, I made him stop. The reasons for the awful noise were twofold: The driver failed to shift to a lower gear whenever the car was engulfed in the slow-moving crowd on Kunming's streets, paced by water buffaloes and pigs plodding in front of the Buick; and the car's engine had, therefore, to labor heavily, with its ignition timing set for American high-octane gasoline and not for the lower-than-normal octane rating of the deteriorated gas stored in China for years. Such a combination could not help but produce a dreadful engine pounding. Rather than criticize the chauffeur's driving and making him lose face, I did the only practical thing: I retarded the Buick's ignition drastically. By merely loosening a single screw, rotating the ignition distributor timing adjustment backward and tightening the screw again, the problem was solved instantaneously! The Buick ran without any engine noise! (Admittedly, the big car now had much less power, but who cared?)

You should have seen the chief's face! I was the Man of the Year, his miracle man! When we returned to his palace, I received a royal payment indeed. The chief had watched me glance admiringly at his Peugeot convertible. He told me, through his interpreter, that he would never use that small car now that his Buick Century was in order again. What better way for him to express his gratitude than to present me with the new green Peugeot? Naturally, I refused, but to no avail. (Any further argument on my part would be a discourtesy to the chief, the interpreter volunteered on his own.) So now I suddenly owned a car in China! Working as a grease monkey, I had progressed from owning a motorcycle in Germany to a leather-upholstered convertible in China's interior. But that was not all: A brown envelope was handed to me by the gatekeeper when I left. In the envelope were three brand new U.S. hundred-dollar bills, twice what I earned during a month at Tessier's garage. Proudly I drove back to the Model Village, in the first car I ever owned.

Next morning another message came from the police chief, written in perfect English by his interpreter. Since I was so successful in reviving his Buick, how about attending to Governor Lung's seven-car fleet used by him, his four wives (the first wife always remained *the* #1, the other three were younger additions living with the husband one at a time—as differentiated from the Moslem custom of having several wives living under one roof) and two of his sons, aged twelve and ten. Governor Lung's cars were American; the cars of his sons were small British Austins, whose foot pedal height of brake, clutch and accelerator I increased by attaching wooden blocks. (The youngsters, incidentally, drove very well.) Furthermore, Chief Li pleaded with me to attend to his Buick Century henceforth on a monthly basis, at his own garage.

There was nothing that could possibly go wrong with a brand-new Buick that was driven at a top speed of 20 miles per hour, however poorly, only 300 to 400 miles per month during the daily air raids. Again, I had to be careful not to make the chief lose face. After all, he *had* given me a grand present! From then on I drove my Peugeot to the chief's pagoda every month. When I arrived on the dot at 9 A.M. (the Japanese switched their bombing schedule in autumn 1941 from 10 A.M. to later in the day), I found the Buick already jacked up and all four wheels placed on wooden blocks. A red rug had been spread underneath the car for my comfort; officers stood by to assist me if it was necessary (which, of course, it never was). Chief Li had his large bamboo fan chair placed about 100 feet from his Buick's open stall, to observe what I was doing and to make sure that I would receive the proper care he had prescribed to his staff when my work was completed.

The procedure was identical at each visit: After our formal greeting, I crawled underneath his Buick. There was little I could do to it without doing more harm than good. (I had learned early during my apprenticeship: If it isn't broken, don't fix it!) So, I cleaned the fuel filter bowl; every three months changed the engine oil, which did get somewhat black with all that dust on the road; sometimes I looked at the Buick's spark plugs and ignition breaker points, cleaned the carburetor air filter, and checked the car's brakes. Once I refilled the shock absorbers. Whenever I asked the interpreter to tell the chief that I could find nothing wrong and that we should reduce the frequency of my visits, he would hear nothing of it. (Obviously, the chief had heard about preventive maintenance.) After fiddling around for about an hour, enough for my Chinese host to know that I was "doing something" to his Buick to make it 1,000 percent reliable, I slid out from underneath the car. An officer in tailored uniform poured warm water into a white enamel basin which another officer held; yet another brought soap and held a towel ready for me to use. I washed my hands and face in front of everyone including the chief's household personnel, who formed a large semicircle around Chief Li, at a respectful distance.

Then came the invariable question from the chief via the interpreter: Would I honor Li Tse Cho with my presence at lunch? The answer was foregone, of course. At each such monthly luncheon were present the commanding general of Yunnan's air-defense system and the mayor of Kunming, both already customers of mine. Food was served at the round table in a large bare dining room, course after course; the host traditionally dished out the first morsel of each new delicacy, placing it with his own chopsticks (which he had been using for himself during the meal) into the bowl of the honored guest. Rice and soup closed the lunch. It was natural for host and guests to spit onto the floor chicken, duck or pigeon bones, which were promptly eaten by his two dogs, waiting under the table. Few words were spoken during a meal; one concentrated on what was being served.

At one lunch the rice served was spicy. I didn't—and couldn't—know this. Using my chopsticks properly and holding the bowl to my lip, I shoveled globs of rice into my open mouth, filling it to capacity. Only then did I find out that this particular rice was too spicy for me to swallow. I tried to avoid coughing; by that time my eyes were tearing. I began to choke; with an eruption, I exploded. Rice kernels were flying all over the table! I apologized, but the reverse happened: "With the deepest of regrets, the chief asks you to accept his apologies for having served such terrible food. . . . Would you forgive him, please?" asked the interpreter. One just can't beat old-fashioned Chinese courtesy.

It was a cool morning when December 8, 1941, came around. (Remember the Date Line day change across the mid-Pacific?) I left my house to go to work as usual, at 7 A.M. . . .

7

"Herman the German" in the Flying Tigers

It was October 1944. "And what in hell is wrong with getting him into the States illegally?" thundered the General at his aide, Colonel Strickland. Claire Lee Chennault, former Commander of the original "Flying Tigers" of the Chinese Air Force, now promoted to Commanding General of the United States 14th Army Air Force and renowned the world over, was the first man to clobber the Japanese Imperial Air Force at the beginning of America's entry into World War II. The General crouched behind his wooden desk in his simple headquarters. He slammed his fist down so hard on his desk that the tea spilled out of his cup. "This man has been doing a great job for us!"

I was standing a few feet from the chief, still in my splattered flying suit after a thirty-six-hour, nonstop 550-mile Jeep ride on a slippery mud and red-clay road from the southern air base at Liuchow. Chennault had called me to his headquarters in Kunming to hear my latest observations on the eastern front; he planned to send me to Washington for a face-to-face report to the Commanding General of the OSS (Office of Strategic Services, the wartime military intelligence agency). He also wanted me to get a field commission and become an officer since I handled classified material. Chief of the OSS was General "Wild Bill" Donovan, World War I hero and winner of the Congressional Medal of Honor. Here was my second chance to visit America; eighteen months earlier, a bureaucrat from CBI (China–Burma–India) headquarters at New Delhi, India, had nixed the first order from the Commanding General of all U.S. Air Forces, four-star general "Hap" Arnold, for me to accompany my Japanese Zero fighter plane to America and help sell war bonds. That low-ranking paper-pusher in New Delhi had dug up my 1942 enlistment documents and read the small print at the bottom of one of the pages: "Eligible for duty outside the continental limits of the United States and at the War Zones in the Pacific only." So after Chennault had been requested by the China–Burma–India headquarters to forward my personnel folder to the Pentagon a year and a half ago, that bureaucrat had stopped my first visit to the USA before I left India in March 1943. Now General Chennault was trying again.

Poor Colonel Strickland—he knew very well what so infuriated the General. Strickland was familiar with my case and the Adjutant General's ruling that supported that bureaucrat's objection; hesitatingly, he told General Chennault, "Sergeant Neumann's trip to the USA would be illegal, sir."

Immediately following his desk-pounding, Chennault asked rhetorically, "Who in hell is running this war anyway, the State Department or the Army?" He then instructed Colonel Strickland to disregard the red tape and to have travel orders typed for me at once, sending me to Washington "to carry out secret verbal instructions by the Commanding General of the 14th U.S. Army Air Force." Thus no one could ask me any questions en route.

"Bring the orders in here," called the General after the colonel. "The sergeant will wait for them." He winked an eye at me after Strickland had left. "That ought to do it! Take the next plane out of here. Get going before some paper jerk creates more problems." With that he reached into his top drawer, pulled out a three-by-four pen sketch of his strong face and wrote on it "Best Wishes, Neuman, C. L. Chennault." (His picture stands on my desk in front of me.) Ten minutes later Colonel Strickland was back and handed my travel orders to the General. Chennault signed them, then called after me, "Thank you for everything, Sergeant!"

The incredibly lucky three-year path that led to my setting foot on American soil for the first time as a U.S. Air Force master sergeant who was still a German citizen had begun in China on December 8, 1941 (Far East time). It was a sunny, wintry morning in Kunming. I was just leaving my house to walk to my garage when my neighbor, Claire Lee Chennault, was getting into his old Ford. He stopped me. "Have you heard the news about the Japanese air attack on the U.S. fleet at Pearl Harbor?" I had not; I didn't even know where Pearl Harbor was. Chennault told me briefly what had happened that Sunday morning in Hawaii. He confirmed the open secret that one of three pursuit squadrons of American P-40 planes now in Burma, flown by Americans, would land in Kunming in a few days and that he was their commanding officer. His rank, given to him by the Chinese Air Force, was colonel.

Chennault asked if I would now give up my garage and join his ex-Army, Navy and Marine pilots and mechanics whose formal four- to six-year contracts with the U.S. Government had been prematurely terminated, on condition that these men would serve with the Chinese Air Force in a clandestine U.S. military operation. Three squadrons of Americans, all volunteers, had been retraining intensively—according to the rules Colonel Chennault had laid down—north of Rangoon since mid-'41, while Curtiss P-40 "Tomahawk" airplanes for their use were still being assembled by CAMCO (Central Aircraft Manufacturing Company) near Lashio in Burma (on the British side of the Burmese-Chinese frontier, to be safe from Japanese air attacks.) Two hundred fifty-two men and two nurses formed the first "American Volunteer Group of the Chinese Air Force," the AVG. Long-range plans for two additional groups to follow the first AVG into China were formed secretly in Washington, under the guidance of shrewd Tom Corcoran. He was a trusted friend of President Franklin D. Roosevelt and one of his famed "Whiz Kids." In prior years there had been an Italian fighter squadron, later two Russian fighter and bomber groups, helping the Chinese rather ineffectively because of their outdated Spanish Civil War equipment.

I made up my mind on the spot: "Yes. I'll join you." While such a hasty decision seen through the eyes of age, wisdom and experience seems rash, my reply was based on the confidence I had in Chennault, my neighbor of seventeen months. As it turned out, it was the best rash decision I ever made. I turned over my Reliance Auto Service to a German refugee who had joined me two months earlier, returned the lovely green Peugeot convertible to the chief of police, and moved into "Hostel I" with the first two squadrons of American military personnel in Nationalist China.

There were about fifty men in that contingent, part of those volunteers who had left the USA for Burma six months before Pearl Harbor in the guise of tourists. Their professions were listed on their travel passports as anything from musician to bank vice president. The men and two nurses had traveled on the Dutch excursion liner *Jaeggers Fontaine* dressed in sport shirts, with straw hats and cameras—all with the specific approval of President Roosevelt (and, I was told, unbeknownst to Congress). What was my impression of the first group of Americans whom I met in Kunming? I liked them. All were nice to me. There was a marked difference between most of the college-educated officers and many of the enlisted men, noticeable in their language and behavior toward the Chinese people, girls in particular. In general, the men who had enlisted in the U.S. Army years before were hard-drinking, poker-playing, rough and tough, yet they called each other by first names, showed me photos they carried in their wallets of "Mom and Sis," "Dad" and their dog. They went out of their way to teach me the American way of life. Never before had I heard of Log Cabin syrup, hotcakes or waffles. I didn't know what to expect when one of the Americans called at the breakfast table, "Shoot the jam, Sam!" I had never seen—or even heard of—a baseball game, or a football game. When one of the men and I walked behind an attractive Chinese girl in a tight silk dress, my Texan friend sighed. ". . . I'd like to bite her in the ass and let her drag me to death!" I couldn't imagine what this Texan had in mind: I knew neither that three-letter word for posterior nor why he wanted to be dragged to death. But I understood when they gave me the nickname "Herman." "Herman the German" stuck with me throughout the war and, in some cases, even today! (At one mail call in 1944 I received a postcard from a buddy from India addressed simply "To Herman. U.S. Air Force. China.") In contrast to the enlisted men, most AVG officers and pilots were reserved, quiet and very courteous. They kept their distance from ground personnel. There was no formal military protocol or dress code amongst AVGers; according to Chennault, we were in China to fight the Japanese, not to play soldier and worry about making up our bunks so tight that the proverbial penny bounced on the blanket.

Life for the inhabitants of Kunming changed dramatically for the better after Pearl Harbor and after the mid-December '41 arrival of the Americans to protect their town against air attacks. During their first encounter at Kunming with U.S. fighter planes awaiting them at altitude (thanks to the Chinese ground warning net, which alerted the Kunming-based squadron an hour before the Japanese arrived and gave the Americans time to get to the high altitude), the surprised Japanese—unescorted by fighters—lost five of their nine bombers. The others must have been severely damaged and possibly they crashed en route home, since a reliable intelligence source from Hanoi reported that only a single bomber returned from that raid. Five days later, they made one more raid with nine bombers and lost, once again, five planes. The surprisingly slow and cumbersome Japanese bombers had to execute a 180-degree turn (a maneuver they carried out flying in formation!) to return to their base in Indo-China. They were extremely vulnerable as they banked to make the turn and presented in that position a perfect target for our fast-firing, slow-flying P-40's—hitting the bombers' unprotected bellies and wings. That was the last air raid over Kunming for more than two years. No more daily bombing raids by training planes flying three V formations, dropping their loads of five-hundred- and thousand-pound explosives unmolested into the "open city"!

Just before Rangoon fell to the Japanese ground forces in March 1942, the last of the three American Volunteer Group squadrons —having already lost some men and planes to enemy action and to accidents—transferred from Burma into China and was dispersed along the eastern and southern fronts; each of the three squadrons was frequently shifted from one base to another to deter the Japanese from attacking with impunity any area in Free China. Our planes had sharks' teeth painted below the propeller spinners, identical to certain German Messerschmitt Me 109 fighter squadrons. The Flying Tiger insignia, a tiger with little blue wings jumping through a large *V* (for victory), was a creation of Walt Disney. AVGers quickly became heroes in China and the United States; not only because of their dramatic shoot-down ratio of enemy aircraft to their own losses (14 to 1) but also because the Allies badly needed encouraging war news in the first few disastrous months of 1942. AVG pilots had one advantage over the Japanese: They

During the last eight days in May 1939, I traveled from London to Hong Kong in an Air France Dewoitine 338. Only nine were built.

Typical section of the original Burma Road over which I took a Chinese convoy of twenty Renault diesel trucks, 580 miles each way to the Burmese border and back to Kunming, in 1940

Tenth-century water buffalo carts lurch past a twentieth-century Flying Tiger P-40.

American Volunteer Group P-40's of the Chinese Air Force in 1941

謹向飛將軍致敬！

中國空中的保衛者——中國神鷹隊的英雄們，美國飛虎隊的英雄們：

在我們悠久的渴望中，在我們高度的興奮中，六月十二日你們在桂林的天空，追擊和殲滅了日本的荒鷲八架，創造了桂林空戰史上空前的光輝戰績，並且又一次創造了你們空中殲敵的偉大榮譽。這在我們桂林三十萬市民中，無不特別感到興奮與歡快！同時，在中美兩國共同殲滅一切侵略暴力的合作上，更是一種無上的光榮！深信全中國與所有盟國人民的內心，必能永久留着不可磨滅的紀念。

今天，我們桂林三十萬市民謹向你們慰勞，並向你們謹致崇高的敬意！

同時，還望你們更加發揮英勇卓絕的精神，繼續創造更大更豐的戰績！

最後我們高呼：

中國神鷹隊萬歲！

美國飛虎隊萬歲！

中美合作萬歲！

桂林各界慰勞空軍晚會 六月二十日

TO OUR FLYING HEROES

Guardians of the air, you heroes of the American Flying Tigers and the Chinese Divine Hawks:

After our long expectation and to our great cheerfulness, you have annihilated eight Japanese vultures in the air above Kwelin on June 12. This is the most brilliant merit of air combat that has ever been achieved at Kwelin. You have once more created your great glory of extinguishing the enemy in the air. We, the 300,000 citizens in Kwelin, are especially stimulated and delighted! While considering the co-operation between the U. S. A. and China for the common resistance against aggressive violences, this is a particularly incomparable glory. We are sure that all the people of China and the Allies will take this to be an ever lasing token of remembrance.

Today, we, the 300,000 citizens in Kwelin, are presenting you our heartiest congradulations and highest respects to your comfort. And we are expecting your continual achievements of greater and richer merits with your inexhaustible heroism and bravery.

Let us yell:

Long live the American Flying Tigers!

Long live the Chinese Divine Hawks!

Long live the co-operation between the U. S. A. and China!

The Kwelin Air-men Comforting Evening Party

June 20, 1942.

Here is a battered copy of the invitation to the "Comforting Evening Party" honoring the original Flying Tigers.

Technical Sergeant Neumann in the cockpit of a Curtiss P-40

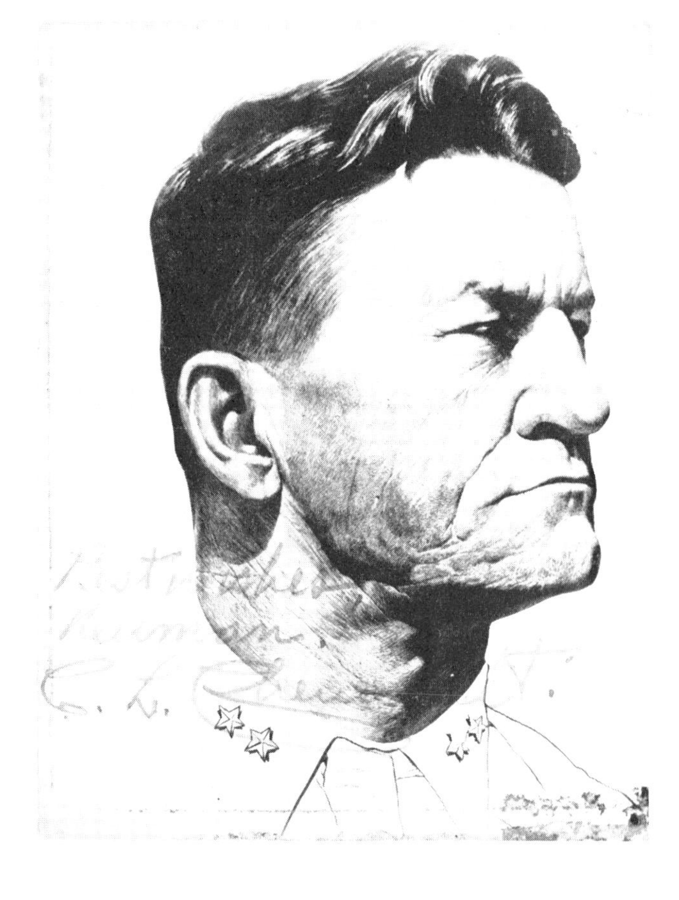

To the Flying Tigers, and later the 14th Air Force of the Army Air Corps, I was always "Herman the German." Only General Chennault called me Neumann.

My reconstructed Japanese Zero after the landing accident *(above)*, and after I rebuilt it again

One of our OSS radio Jeeps (before we had to blow it up) which permitted us to direct, from camouflaged positions, attacks by our low-flying bombers

With an escort of Chinese soldiers, on one of my first OSS missions within enemy lines to gather intelligence

Here I have a captured Japanese flag and helmet in September 1944, after an OSS mission with Lieutenant John Birch.

The Chinese riverboat on which I hid while floating through Japanese lines between Wuchow and Canton

G.I. IN A JAM

Gerhard Neumann lost his German citizenship by refusing to return home when war started. Remaining in China, he joined the Flying Tigers in December 1941, later enlisted in the 14th AAF. Now recommended for a commission, M/Sgt. Neumann must first obtain U.S. citizenship. Under service naturalization rules he may have to be discharged, re-enter the U.S. as an immigrant, reenlist to qualify.

The Overseas Edition of *Life* magazine, November 13, 1944, featured De Gaulle on the front and "G.I. IN A JAM" on the back cover.

could bail out of a fatally crippled plane over "home territory" and were assured of a most friendly reception by a hospitable Chinese population, which returned the pilots to their home bases promptly, in spite of a $10,000 cash premium on the head of each American flyer turned over to the Japanese. Japanese pilots did not even bother to wear parachutes. None of them would have stayed alive for long anyhow if the Chinese captured them.

The Japanese learned quickly that air warfare could be more than a one-sided affair and adjusted to the AVG's presence. Whenever they did attack our bases thereafter, it was only with great numerical superiority. After their December 1941 losses over Kunming, Japanese bombers never again came unprotected during daytime raids over American air bases but were escorted by Zero fighters. In contrast to their inferior bombers, the Zero was superb. It was armed with two .30-caliber machine guns firing through its propeller and two 20-millimeter cannons mounted in its wings. The Japanese plane was much more maneuverable, further ranging, faster climbing and higher flying than our twice-as-heavy American fighter planes in China. The Zero had an engine with horsepower equal to ours yet the plane weighed only half as much; consequently, our Tomahawk P-40's, designed in 1937, were of dramatically inferior performance. The rugged and heavy P-40 could be shot full of holes like a Swiss cheese and still keep on flying. The Zero, in contrast, exploded in a ball of fire when hit by American incendiary and explosive shells from fast-firing .50-caliber machine guns.

To fool the Japanese about the small number of planes the AVG had available to protect the still immense area of Free China on any particular day, Chinese carpenters did a marvelous job of building P-40 aircraft out of bamboo and wood. The dummy planes were painted with squadron insignia including the Flying Tiger sharks' teeth. They looked so realistic that one of our pilots, somewhat inebriated, mistook a dummy for his real plane during an alert and tried to climb from the wooden wing into its cockpit! At night the dummies were moved about so that the Japanese "Photo-Joe," as we called their reconnaissance plane flying daily over our bases every morning at very high altitude, was constantly misled about the actual number of planes we had.

By mid-'42 the Japanese Air Force had learned to respect the

Flying Tigers as "honorable enemies." Leaflets were dropped over Kweilin Airfield by a lone plane: "We, the flyers of the Imperial Army, challenge you, the flyers of the Flying Tigers, to a sportsmanlike duel over Kweilin on June 28, at 3:00 P.M." All of us suspected a trick or a trap, except Claire Chennault, who knew the Japanese well. He ordered a squadron from another base to join the one already at Kweilin on June 28; four P-40's were sent aloft in a southwesterly direction and ordered to circle at as high an altitude as they possibly could, keeping the sun at their backs, ten minutes before the "appointment." The remaining twenty-six P-40's of the two squadrons waited at the end of the runway, their engines warmed up and idling. Chennault gave orders for their go-ahead fifteen minutes before 3 P.M. Right on time, shortly before three o'clock, the distinct drone of a large fleet of Zeros could be heard. Barely a minute after ground radio operation alerted our four planes high overhead that sixty Imperial challengers were approaching, the two dozen airborne P-40's were climbing at maximum speed.

Our four circling planes came out of the sun, pouncing on the Zeros with a deadly stream of rapid-firing machine guns. The first half-dozen Zeros, blinded by the sun and unable to see the attackers, began to smoke, spiral to earth or disintegrate in the air. Our P-40's had been ordered not to get cornered by the overwhelming number of Japanese fighters, and not to be ashamed to dive away from a dogfight. At that time we suspected—but did not know—that Japanese Zeros were unable to dive fast. The battle took place right above our Kweilin airfield, planes zipping about in all directions. It was a fascinating spectacle. Ten minutes later—it seemed like an hour—fuel supply on each side was getting low, and the enemy withdrew. They had left behind fourteen planes and pilots; four of our own men had had to bail out. Not one American pilot was lost or hurt.

Toward the end of June 1942, we were informed that the American Volunteer Group was to be disbanded soon and that the regular Army Air Corps would take over. On July 3, one day before the AVG was dissolved, all members of the original Flying Tigers in Kweilin—by now of world fame—were invited to an evening party given by the local business community. A bevy of beautiful, slim, English-speaking girls in slinky Chinese silk gowns and a flower in their black hair, escapees from Shanghai and Hong Kong, were waiting for us on the sidewalk. As we climbed off the trucks that brought us to town from the airfield ten miles away, the girls took each of us by the arm and escorted us, one by one, into the large dining room to the applause of our hosts. The grateful merchants had organized what they called a "Comforting Party." It became one indeed, lasting late into the night. . . .

Why were the Flying Tigers disbanded if they were doing so well? Because Chennault was boycotted by the U.S. War Department. No replacement aircraft, no ammunition and no spare parts arrived for him in China to support his tiny volunteer group of 253 men and two nurses. Many Air Corps generals in the States were jealous of the former schoolteacher and retired Army Air Corps captain who had become a world hero. Finally—and this I agreed with—it was not fair to have American ex-servicemen fighting side by side with regular U.S. Air Corps pilots, all facing the same odds in the war, yet one small group receiving a multiple salary from the Chinese government plus a cash bonus for each Japanese plane shot down.

Retired American Captain, Chinese Air Force Colonel Chennault was persuaded to reenter the U.S. Army Air Corps in China. President Roosevelt offered him an immediate four-level promotion to brigadier general, which he accepted. Thoughtful Chennault, without telling me in advance, had requested (and been granted) permission from Secretary of War Henry Stimson to enlist me in the U.S. Army Air Corps at the time of the disbandment of the American Volunteer Group. AVGers who volunteered to remain with Chennault in China beyond July 4, 1942, and whose war experience was immensely valuable to the first contingent of freshly arriving "green" U.S. Air Corps personnel, were offered promotional incentives, whether enlisted men or officers. Despite such incentives, only 35 of the 255 stayed on.

The reason why so few AVGers stayed with Chennault was threats of an "overseas draft," uttered in undiplomatic language to the assembled Volunteers the evening before the official takeover of the AVG by the Army Air Corps. (Americans could not be drafted unless they were on U.S. soil, according to the law of the land.) Brigadier General Clayton Bissell had been dispatched to

China by the War Department for the specific purpose of persuading as many AVG personnel as possible to reenlist with Chennault, I was told. These men loved our "Old Man" but wanted no part of Bissell, who implied that he would have the overall command. Instead, they taught small Chinese boys to stand at the side of any road, point their middle finger into the air, and yell as loudly as they could at passing Americans, "Piss on Bissell! Piss on Bissell!"—without knowing what these English words meant, of course. The AVG-created endearments for General Bissell spread like wildfire around the Kunming area and did not die down for a couple of years. A few American Volunteers returned to the States, where they were drafted promptly; the great majority accepted much higher-paying civilian job offers overseas, flying for airlines under contract with the U.S. defense establishment or working for American manufacturers, assembling tanks, Jeeps and trucks to be shipped as lend-lease aid to Russia from the coast of southern Iran.

Farewell ceremonies were scheduled for the departing original Flying Tigers. Fox Movie News was set up to film the historic event. Those few airmen remaining in China, reinforced by men who offered to stay for two extra weeks to help General Chennault during the transition period, were to form three ranks. Two P-40 fighter planes with sharks' teeth formed a suitable background for a newsreel. Although I never had any U.S. Army basic training, I was persuaded to participate to help swell the ranks.

My buddies and I knew that American military commands were nearly identical to the German, and so were their executions. To be safe, I was to place myself squarely in the middle of the small group. "Don't worry!" I was told. "The man next to you will tell you what maneuver is expected after the first part of the command has been called out." (Each military drill command has two parts: The first one tells the soldier what the maneuver is to be; and the second one, called out seconds later, that it is to be carried out.) I was standing, somewhat nervously, in the midst of my buddies when an order came to "Spread ranks!" meaning the participants were to move forward, backward, to the right and to the left. This caused me to stand all alone smack in the middle of the group. So far, so good! Then came the command "About———," i.e., the order to turn 180 degrees. My neighbor alerted me to turn around when the word "face!" would be called out three seconds later. It was, and I turned around.

But there is a major difference between the American and the German execution of the command. I had already completed three-quarters of my 180-degree turn to the left when I noticed everyone else using the American system; that is, turning to the right. Instead of leaving bad enough alone, I rotated back on my left heel—then did it the American way! Fox Movies had to cut out several feet of their valuable news film. It was the first and last American military close-order drill in which I ever participated.

On the evening of July 4, 1942, in the dim, oil-lamp-lit room of Group Surgeon Dr. Gentry in the Kweilin barracks, I was sworn into the U.S. Army as a staff sergeant although theoretically still a German national. I was given an embossed, small metalic plate, worn around each soldier's neck for identification purposes; it was called a dog tag. Mine read ASN 10500000. They told me that 1 stood for "enlisted," the first 0 for *no* district (America was divided into nine districts identified by numbers 1 through 9) and that the 5 in my number was inserted merely to make an otherwise incredible Army Serial Number 10000000 a bit more believable! I signed two papers without reading the small print. Thus I was unaware that one paper stated I was to serve on the Japanese front exclusively, to avoid any possible capture by German forces. The other stated that I would be ineligible—as an enemy alien—to be commissioned an officer, to be decorated (in contrast to treatment accorded cobelligerents and neutrals) or even to enter the United States! The week after enlistment, I was promoted to technical sergeant; later, I was made the squadron's master sergeant, the highest rank for a non-commissioned officer.

A few weeks later, after our Air Corps had replaced the Volunteers, a lone shot-down Japanese flier surprised us by using a parachute. He became a prisoner of the Chinese, who turned him over to us for interrogation, a Western meal and a few cigarettes. He was safe and well while in our hands; once we returned him to his Chinese captors, his head—neatly separated from the rest of his body—was stuck onto a long bamboo pole and paraded around Kweilin city before being implanted at the entrance to our airfield.

I began my service in the U.S. Army Air Corps as engineering chief of the 76th Fighter Squadron, one of the three squadrons which made up the 23rd Fighter Group. This Group was built around the thirty-five officers and men of the original Flying Tigers.

The first group commander, who was not a member of the AVG, was Colonel Robert Lee Scott (who later wrote *God Is My Co-Pilot*), reporting to General Chennault. Supplies remained minimal even after we became a unit in the U.S. Air Corps. (The North African campaign shortstopped many of the goods destined for China.) The Burma Road had been sealed off by the Japanese as soon as they reached its western terminal. China, in 1942, was thus cut off from the world except for shipments flown in from Assam (in eastern India) over the 20,000-foot-high, snow-covered hills of the lower Himalayas, with many uncharted peaks higher than 23,000 feet. A number of planes and their crews were never seen again after they had taken off from India's Assam for China; they had run into the nearly permanently bad weather, were unable to see, and crashed into mountain peaks.

My first assignment in the U.S. Air Corps' China Air Task Force was to recover three ex-AVG P-40 fighter planes left behind at the evacuated forward base at Kweilin; the planes needed repairs before they could be flown to the rear. A staff sergeant and I worked from dawn to dusk, seven days a week, to make them flightworthy. The last evening, the Chinese base commander gave a small party for the two of us, to which he invited a charming young Korean, Angela A. During the past three years she had fled from Seoul, then Shanghai and lastly from Hong Kong, always ahead of the Japanese troops. Like many other refugees, Angela had walked the last two hundred miles across rice paddies to reach safety at an advanced American air base. Her father had been the Korean Ambassador to Berlin for five years, and her uncle—Prime Minister of Korea —had been assassinated by the Japanese in 1924. Angela, who spoke German and English fluently and with whom I became close friends, was to save my life a few months later. The sergeant and I left the day after the party for our main base in Kunming, Yunnan.

After learning from me about my father's death, the Chinese government thoughtfully sent a cable to my mother via its embassy in Berlin: "The Government of China would be honored to welcome the mother of a man helping China." She secured a German travel passport, flew with Lufthansa from Berlin to Moscow in October 1940 and continued by train to Alma-Ata along the Russian-Chinese border. She connected there with the German airline Eurasia (a Lufthansa subsidiary) for her flight to Kunming, where she arrived elegant as usual and fresh as a daisy—after an overnight stop in Hami, near Mongolia.

Former neighbor Claire Cennault did not think it safe for my mother to remain alone in war-torn China after I had joined the U.S. Air Corps. He suggested to me that I persuade her to fly over the Hump, the foothills of the Himalayas, in a transport plane and then take a BOAC flying boat from Bombay to Cairo. From there, it was only a few hours by train to Jerusalem, where she joined my sister in 1943. After the War of Independence in Israel in 1946, she traveled to America and became a U.S. citizen. Until her death, she traveled extensively in Europe and the Mideast.

8

Unraveling the Mysteries of the Japanese Zero Fighter

One day in October of 1942, when not much was expected to happen, I was standing with a group of newly arrived GI's on the wing of a P-40 near my squadron's hangar at Kunming. I was explaining "real world" maintenance of a fighter plane engaged in action compared to theoretical instructions given in the States. Tall Lieutenant Colonel Bruce K. Holloway (after the war to become Vice-Chief of the U.S. Air Force, and four-star Commanding General of the Strategic Air Command) strode over to the fighter plane. "Herman, the Old Man wants to see you right away!"

General Chennault's office was three minutes away. "Neumann," said the General, "we've got hold of a pretty good Zero captured by Chinese farmers on the Japanese-occupied beach opposite Hainan. They took it apart and dragged the pieces inland. Whatever they've got has supposedly been deposited this side of the Japanese lines, but probably much is damaged or has been lost. There must be useful parts in the wrecks of the many Zeros we've shot down over the past nine months. Here is a marked-up chart where the wrecks are located. How about trying to put *one* Zero together, to test-fly it against our own planes?"

I was speechless. A jumble of thoughts raced through my mind: German, noncommissioned officer, the most famous, mysterious and admired Japanese plane, the *Zero*! Before I could answer, the Old Man added, "It's vital to America to put one of those damn planes into flying condition. If anyone can do it, I'm sure it's you!"

"Thank you. Yes, sir!" I was completely overwhelmed.

Chennault handed me a one-page pass in English and Chinese which had already been prepared with my photo and his signature, requiring only my thumbprint. It gave me the highest priority for anything I might need, be it transportation, the right to interrogate or take into custody Japanese prisoners, or any assistance from Chinese authorities. To assemble the Zero, the first in American hands, without any drawings or manuals whatsoever, was a fun challenge for any engineer. I selected Texan Sergeant Mackey (the GI who wanted to be dragged to death by the Chinese dame a few months earlier) to assist me. Half a dozen Chinese mechanics were supposedly awaiting us at the location where the pieces of the Zero had been deposited. This project, classified secret, was to be done in a wooded area a few feet away from an emergency fighter strip, very close to the Japanese lines. Of course, there was neither hangar nor electric power; all we had to work with were wrenches, hand drills, an assortment of rivets and hammers, and a charcoal fire to heat and straighten out bent propeller blades. We placed the fuselage with both wings on three empty fifty-gallon gasoline barrels. It became obvious immediately from parts we had removed from other Zeros, which had spun into rice paddies after they were shot down, that these planes had not been mass-produced—an important fact for our intelligence operation. Most components required filing before they finally all fit together. We were unable to recover any original tires—which the Chinese farmers had cut up for shoe soles; the tires we obtained for our Zero came from an ancient American Hawk biplane. Japanese fabric used to cover the Zero's ailerons, rudder and elevators had been converted into clothing, so we used Chinese silk with several coats of paint.

Japanese design philosophy was far advanced, intelligent and interesting. Like today's automobiles imported from the Land of the Rising Sun, the Zero included many creative innovations. Its oil cooler, oil tank, American-type propeller and engine were assembled into one single unit held with only four big nuts to the plane's firewall (the front of a fuselage). All fuel, oil, pressure, temperature and other service lines were connected into a single, simple, "quick-disconnect" junction box. To install or remove a complete Japanese power plant together with its propeller and oil cooler system took twenty-five to thirty minutes, while a similar job on its American counterpart, the Curtiss P-40 or newer North American P-51, needed five to eight hours! Such dramatic time advantage meant very much in a combat area, where an extra plane in the air could mean a victory while one left on the ground could become a loss due to strafing or bombing by the enemy. The Zero's right and left wings were one integral assembly with the cockpit, to save the weight of flanges and bolts. Its landing gear was light, weighing one-third of a P-40's. Other components of the Zero such as the gunsight or oxygen system were similar to or copies of German and Russian designs.

We were certain that Japanese intelligence was alerted to our efforts to put one of their Zeros into the air, and that they would try to prevent us from learning its secrets. Their Photo-Joe circled our work area daily to discover where we hid the plane, but Chinese soldiers, unsurpassed in the art of camouflage, thwarted his efforts. It took two months before I sent a coded radio message to our headquarters that the plane was flyable. We were anxious to move it away from the nearby Japanese lines and back to our own advanced fighter base at Kweilin as quickly as possible. We filled the plane's tanks with 100-octane gasoline, installed a Japanese "belly tank"—an auxiliary fuel tank inserted in the plane's belly, which could be dropped immediately before a fighter plane engaged in combat—and filled it to the brim with seventy-five gallons of fuel. (In retrospect, this was a most stupid thing to do.)

Early next morning, a B-25 bomber landed on the short emergency strip, with Major John Alison aboard to fly "our" Zero for the first time. He was a top-notch pilot who had become the first U.S. Air Corps ace in China. Remember the war song "Johnny Got a Zero"? Four P-40's were circling overhead to protect the takeoff area and to escort the Zero to Kweilin. The B-25 was to take Staff Sergeant Mackey and me to Kweilin also. I taxied the Zero to the end of the gravelly runway strip, then changed seats with Major Alison. I had kept the engine running. Standing on the Zero's wing and leaning over its cockpit, I showed the major where the various engine and flight controls were located and how to interpret Japanese instruments, then wished him good luck. We did not want to spend one minute of unnecessary time at this emergency strip, always fearing that a Japanese bomber might show up at any moment and put a finis to our efforts. I sprinted to the B-25, which had kept its two engines running, climbed up its retractable ladder and crawled into the Plexiglas tail gunner's position. There I lay on my stomach, my camera at the ready. We took off. Circling low over the Zero, we saw that its propeller was not turning. . . . Our B-25 had to land again. I hurried over to the major and explained to him in a bit more detail what lever to move to keep the Zero's engine from stopping again.

With the help of two Chinese mechanics, who turned the hand crank on the side of the plane, I restarted the Zero. I suggested that the major quickly get airborne ahead of us and fly "low and slow" toward our air base at Kweilin, giving our bomber a chance to catch up and pass the fighter. I wanted to take close-up photos of this prized aircraft in flight, then land ahead of the Zero at Kweilin. Major Alison was to fly over the airfield there, at low altitude, so that I could observe the "down" position of the landing gear. If okay, I would give him the prearranged signal to land.

This time the engine kept running as Alison advanced the throttle for takeoff. My muscles tensed, and then . . . what a thrill to see the Zero lift off the ground, a bit wobbly at first as the major got a feel for the sensitivity of the various flight controls! He pulled up the wheels and made one pass over our heads, waggling his wings in the classic pilot's okay. I kept my fingers crossed.

Word of a Zero's flight to Kweilin—flown by an American—had spread. Hundreds of GI's lined both sides of the runway. General Chennault was there; so was U.S. Ambassador to China Patrick Hurley, who had come for this occasion from Chungking. High above us, the four P-40's kept on circling protectively. As agreed, our B-25 landed ahead of the Zero. The plane looked great to me as Alison buzzed the field. The crowd applauded as the major made a low and slow pass. The lowered landing gear looked good also; I cleared him to land by waving a green flag.

Major Alison did a skillful job of piloting this plane, different from anything a U.S. pilot had ever flown. He approached the runway at 60 miles per hour, let the Zero's wheels touch the ground gently. (Since we had been unable to obtain a small Japanese airplane battery, the plane was without radio communication and also without internal warning lights.) With the engine throttled back and the plane still rolling at 30 miles per hour, the Zero's right landing gear suddenly folded inward and the right wing tip hit the gravel runway. The plane spun around, breaking off the left landing gear, twisting propeller blades, wings and the fuselage. The seventy-five-gallon, fully fueled belly tank ripped to shreds. It was an absolute miracle that Major Alison was not hurt, and that the plane did not burst into flames as the 100-octane gasoline spilled over the gravel runway while the plane slithered to a stop in a thick cloud of dust. To this day it's still a mystery to me why we had no explosion, fire or injury.

If you ever want to know how it feels to drop from a jubilant and proud moment into the depths of depression within a split second, just think of me and the Zero's landing accident in China! How *did* I feel? Rotten, believe me! I was ready to take my Colt .45 automatic from its holster and shoot myself. When the dust had settled, General Chennault, Major Alison and I inspected the wreckage. In less than one minute I found the cause for the unfortunate accident: A tiny piece of gravel had been blasted by the propeller during takeoff into the landing gear mechanism and jammed it tightly—preventing it from locking down the gear. (Our Zero was off a Japanese Navy carrier, on whose deck gravel did not exist. Its landing gear tolerances were therefore closer than those of Army Zeros.) General Chennault immediately ordered the damaged plane carried off the runway and camouflaged. And his next order? For *me* to reconstruct the Zero, one more time!!! I never expected such a vote of confidence!

Not even during my toughest apprentice days did I work so hard, so many hours a day. My incentive was not only that the nation badly needed data on this plane's performance on which to base fighting tactics, but, even more, the justification of the General's confidence in me. We had a hangar and electricity for power tools two miles from Kweilin's runway. Because the tail section of the crashed plane was not damaged at all, we unbolted it, loaded it on an open-bed truck, and had a Chinese driver bring it to the camouflaged hangar. Staff Sergeant Mackey and I rode on the truck bed, each holding on to the Zero's light tail so that it could not be blown away by the wind. Unexpectedly, the driver of the truck took a shortcut off the road, trying to pass between two big trees. The space was not quite wide enough for the assembled horizontal stabilizers of the tail. But the driver didn't think of that. . . . When I saw what was bound to happen in a few seconds, I pounded violently on the roof of the driver's cab. Too late! The only non-damaged part, the tail, now also needed repairs.

While Mackey, a few Chinese and I were working on the repair, Kweilin was evacuated by our squadron for the monsoon season of 1942, just as it had been a year before. Sergeant Mackey and I stayed behind, together with a radio operator and several Chinese mechanics with whom we had worked before. With the project nearly completed, I fell ill from overwork and sheer exhaustion. Typhus, malaria (which I had had twice before), and yellow jaundice—all together! I ran a 106-degree fever, shivering and shaking every thirty-six hours. No American medic was nearby. That's when my Korean friend Angela came to the rescue. She arranged for the local Chinese hospital to pick me up, and to rub my body with fast-evaporating aviation gasoline to bring the fever down. The American radio operator let Angela talk with General Chennault's headquarters personally. She pleaded for immediate help. Within less than eighteen hours I lay on a stretcher aboard a specially dispatched C-47 cargo plane, with doctor, nurse, chaplain (!) and Angela, whom the General asked to accompany me to Kunming's station hospital.

The doctors there put me at once into isolation. The only visitors were Angela and the Commanding General himself. Chennault ordered the head doctor to let me out as soon as *I* felt well enough to return to the Zero, whose reconstruction was halted pending my return, regardless of the doctor's opinion. Three weeks later I said, "Okay, Doc, let's go!" A wild argument ensued between myself and the colonel, who felt that I was nowhere ready to leave. I won; but the doctor turned out to be right: The plane was barely in the air when I ran out of breath. An emergency landing had to be made at the nearest airfield, one hundred miles away from Kunming. Two days later I was ready to go again. This time the medics made me put on an oxygen mask even though we flew to Kweilin at less than 10,000 feet altitude.

Two more weeks of intensive work, which, wrapped in blankets and a fur-lined flying suit, I directed from a cot placed next to the Zero. The plane was once again ready to make another flight, this time from Kweilin to the Air Corps headquarters at Kunming. Its pilot was my squadron commander, Major Grant Mahony (later killed in the Philippines). To be sure that another landing gear accident would not happen, we agreed to leave the wheels in a "locked down" position. The 450-air-mile flight was completed without any difficulty. Military police were placed around the Japanese plane to protect it, not from sabotage but from GI souvenir hunters! I had to remain in Kweilin until I got rid of my malaria fever. A telegram arrived from the commander of my fighter group: "ZERO OK PERIOD. WILL NOT REPEAT NOT BE FLOWN UNTIL YOUR RETURN. HOLLOWAY." Back in Kunming, I once again checked all aircraft systems before we began the comparison trials and simulated dogfights between our Zero and various types of U.S. aircraft.

In a nutshell: The Zero's half weight gave it the tremendous flight performance advantage we had observed in each aerial encounter. What made it so light? Besides the weight-saving features I mentioned before, such as constructing both wings and the cockpit in one unit, the Japanese Army had eliminated all armor protection for the pilot, omitted inch-thick self-sealing fuel tanks (self-sealing material is useful in case of bullet puncture but reduces the amount of fuel that can be carried and thus the plane's range). They also did away with an electric engine starter, which not only weighed ten pounds, but also necessitated a heavy battery and thick copper wiring to drive the starter motor (plenty of manpower to turn a hand crank was always available on a Navy carrier); they used lighter aluminum alloys than we did, and applied clever design ideas to every minute detail to save critical weight.

But everything in life has its price: The Zero was so light and thin, and its fabric-covered flight controls so flimsy, that it was unable to dive fast when it had to get away from an American heavyweight fighter that was on its tail. This happened seldom, but it did happen. As a result of our flight test evaluations, the order was issued from headquarters to our pilots in China: "Make one pass at a Zero with all guns blazing. Then get the hell away from him! Dive! Cut out all ideas about heroic dogfighting. . . ." Colonel Holloway asked me to write an intelligence report about the Zero to our technical specialists at Wright Field in Dayton, Ohio, where enemy aircraft data were analyzed. I struggled to write my report in English. It was sent to the States by courier mail. Back came a message: "SERGEANT NEUMANN. PLEASE WRITE YOUR REPORT IN GERMAN. WE CANNOT UNDERSTAND YOUR ENGLISH." All evaluation flights, comparing the Zero with anything flyable we had in China, were completed by March 15, 1943. The U.S. Air Force headquarters in Washington issued orders for the Zero to be shipped to the States; first to fly from city to city and sell war bonds, then to be examined by our stateside technical intelligence organization in Dayton.

The following day, aboard a B-25 bomber, Staff Sergeant Mackey and I flew ahead of the Zero over the Hump (the foothills of the Himalayas) and all across India, from Dinjan (Assam) to Karachi, with stops at Agra and Bhagalpur. We checked—and serviced—the Zero at every stop. Each of four P-40's assigned to protect the Japanese plane developed some problem, one on each segment of the long trip, preventing any of the American planes from reaching Karachi. The Zero had not a single problem! When it arrived at India's west coast port with Major Mahony at the controls all the way from China, it landed without any American escort. People there were amazed to see a Zero sitting in the middle of their runway! (Wouldn't it have been ironical if an American plane had shot it down?) The Zero was to be disassembled at Karachi and then shipped by freighter to the USA via Australia. Within hours Major General Bissell, the 10th Air Force Commander in India, paid us a visit, offering any support we might need. Three days later his staff car reappeared; the two American MP's guarding the entrance to the giant British R-101 airship hangar in which we disassembled the Zero presented arms. A gray-haired gentleman in a wrinkled windbreaker emerged from the car with Major General Bissell. The older fellow walked around the plane, already partially taken apart and its fuselage resting, once again, on three gasoline barrels; then this old man began to climb onto the left wing and into the cockpit. Warning him about the wobbly position of the Zero and its value to our Air Force, I asked him to get off the plane at once.

I noticed the four little stars on his collar but didn't think much about it. He must be one of the newspapermen collecting souvenirs, I imagined. (Japanese pilots whose bodies I had to dig out of their shot-down planes wore similar small stars on their collars as rank insignia; the American major generals whom I had met, Chennault and Bissell, each had two large stars on their collars.) The gray-headed fellow followed my request, apologized, then introduced himself: "My name is Arnold."

"Glad to meet you," I said. "My name is Neumann."

General Bissell was stunned. He motioned me aside, whispering, "This is our *top* general, Hap Arnold!" Embarrassed, I apologized in turn to the four-star big boss, but there was no need to do so: He was most gracious, grinning from ear to ear because I had not recognized him. He knew about my past and instructed General Bissell to notify General Chennault that he, Arnold, had ordered me to the West Coast to reassemble the Zero when the shipping crates arrived in California or Oregon.

Two weeks later, three large wooden crates, including one very large one for the Zero's cockpit and its two joined wings, had been built by Indian carpenters. I had tagged all important connections to make the reassembly in the States easier. But my trip to America was not to take place. It was thwarted by the U.S. Government bureaucrat in New Delhi who had studied carefully the small print at the bottom of the enlistment papers I had signed on July 4, 1942, prohibiting me from entering "the continental limits of the United States." I was furious that a civilian could overrule a four-star general's order, a thing which would have been unthinkable in Germany. Now, instead of ordering my return to China, General Chennault radioed me to spend a month at Calcutta's base hospital for a thorough R and R (rest and recreation), my first in several years.

I had the nice feeling that I was being appreciated in China; not only by General Chennault, but also by my squadron and group commanders. In an address before an Air Force Association meeting in 1971, reported verbatim in their magazine, four-star General Bruce K. Holloway—Commander of the Strategic Air Command—referred to me during his World War II years in China when he was a lieutenant colonel and my group commander: ". . . having Herman as a line chief was like having Mr. Kettering run the local Chevrolet maintenance shop. . . ." (Charles F. Kettering was the famous engineer, inventor and research consultant at General Motors.)

9

U.S. Air Corps Master Sergeant, Army Serial #10500000

During my stay in Calcutta, I enjoyed the comforts of a real hospital bed with mattress, hot showers and good GI food. I sat in modern cafés in that cosmopolitan city, ate ice cream and watched sacred white cows wander freely in the middle of busy thoroughfares, stopping streetcars and motor vehicles—just as I had seen them do during my flight to China in 1939. The main attraction in Calcutta was the daily morning appearance of American Red Cross girls who wandered through the single-floor, sandbagged and straw-thatched base hospital, dispensing gifts of toothpaste, paperbacks and chewing gum to the patients in the double rows of beds. The good looks of those young American women, in their light-blue starched uniforms, with sparkling white teeth not blackened by betelnuts, were almost too much to take! In China, except for movie scenes projected against a bedsheet hanging over the side of a truck, or a rare one-day visit by stars like sporty Jinx Falkenberg, we had almost forgotten that there existed real women.

I fell head over heels for one Red Cross girl, Virginia, a charming and shapely blonde from California. (We are still friends and write occasionally, forty years later.) Every morning when she entered our ward, all fellow patients let go with a barrage of wolf howls and whistles. I felt deeply offended for Virginia (in Germany, whistling at a girl was considered an insult) and had a violent fight with my next-bed neighbor, also a master sergeant, which resulted in my sporting a black eye and him a broken right arm. The hospital commander made peace between us and explained to me good-naturedly that any sharp-looking American girl would most likely be insulted if she were *not* whistled at by a bunch of admiring GI's!

There was only so much time a man could rest and recreate. I became impatient, checked out of the hospital and hitchhiked—by air, of course—to China for further assignment. During my seven months' absence, all members of my original squadron had been rotated back to the States. I now had to train a new group of draftees in the practice of real-life maintenance of fighter planes under combat conditions.

As engineering chief, I had sixty men in squadron maintenance. Fewer than half of them were "Let's go!" types who cared only about keeping our few planes ready on the front line. The other men bitched about bad working conditions, lack of tools and spare parts, monotonous food and the absence of American girls. I worked primarily with the dedicated fellows. During the two-month rainy season, when we had plenty of time to spare, my squadron commander gave me a paperback edition of Dale Carnegie's *How to Win Friends and Influence People,* in the hope that I would apply Carnegie's suggestions and be able to get productive output from *every* man in my outfit. It took weeks of thumbing in my tattered English-German dictionary to struggle through Carnegie's book, whose prime recommendation was, as I recall, to "compromise" on just about everything for the sake of friendship. I disagreed wholeheartedly with Carnegie, but gave it the good old college try applying his suggestions. By George, it was miraculous how well the "neglected" men of my engineering outfit, those to whom I had paid little attention up to then, suddenly got along with me! I was invited to play poker and blackjack with them; was asked to accompany them on their nightly trips to a nearby town, serving as interpreter for their deals with the local belles. Indeed, I succeeded in winning lots of friends; but I failed totally to influence them to do more and better work. After one month of sincere effort to do what Carnegie suggested, I gave up and returned to my old modus operandi: With fewer but gung-ho fellows, we kept the squadron in top shape. To hell with how to win friends! There was a war on—and the well-being of our planes had highest priority, friends or no friends.

When there were no aircraft activities and everything was shipshape, we read old magazines, which reached China every so often. One advertisement caught my eye: Zippo's. Two Flying Tiger pilots were standing in front of a P-40 with shark's mouth, lighting their cigarettes with a Zippo lighter. The plane's engine showed only five exhaust stacks on its side instead of the six it really had. I wrote Zippo, telling them about their faulty ad, which might reflect on their technical reputation, and promised to tell nobody if the company would send each man in my squadron a lighter! (Matches in China rarely worked and lighters could not be purchased.) The whole thing was intended to be a joke; six weeks later, however, I received a friendly letter from Zippo's president, and a box of lighters for my boys.

To improve the reliability of engine repairs, I asked my squadron mechanics to "volunteer," if this was suggested by me, to test-fly in the single-seat fighter plane whose repair they had just completed. The pilot would sit on the crew chief's lap; neither would be able to sit on a parachute because the cockpit's canopy was not high enough. I had done this myself several times with my tall squadron commander, Major Robert Costello, sitting on my lap, to identify engine problems which showed up only at altitudes of 20,000 feet or higher. The pilots were enthusiastic about my plan. Consequent improvement in quality of workmanship was dramatic. Way past dinnertime, the airfield looked as if it were invaded by glowworms: The twinkling came from flashlights mechanics used to check—once more—the tightness of pipes or connections they had made, in case I might suggest that they "volunteer" to ride in their planes the next day.

To keep our planes in the air, we needed spare parts, but those did not arrive in China in sufficient quantities. So we cannibalized damaged planes and stripped "hangar queens"—planes moved into a hangar and never expected to fly again; their parts were used as spares for other aircraft of the same model—to obtain what we needed. We repaired what we could with baling wire; once I even sawed, then filed an ignition distributor rotor out of an inch-thick slice of tough, non-(electricity)-conducting water-buffalo horn, as replacement for the recently introduced Bakelite part that had developed a hairline crack. That damage permitted high-voltage ignition sparks of 15,000 to 20,000 volts to arc across its crack in thin air at high altitude, causing misfire and loss of power of that engine. It will interest material experts that the buffalo-horn part lasted a year, as long as the airplane itself did. The gun-firing synchronization of one of the P-40B's had gone out of order and the plane's machine guns shot holes neatly through its own propeller blades. For the lack of a spare propeller, one old P-40 flew combat over nine months with three bullet holes, one in each shank of its propeller blades; to continue with flight operation would have been totally condemned by stateside inspectors.

Owing to the lack of spare tires, we had to place bare tail wheels on ammunition boxes at the end of a runway. We told our pilots to run their engines at full power, at the same time to step firmly on the planes' brakes—until the propeller air blast had lifted the planes' tails horizontally. The pilots then let go of the brakes suddenly and took off so that the tail wheels, without tires, never touched the ground. Transport planes, too, had tire shortages: Rubber was badly cut up by sharp-edged pieces of bomb shrapnel that were indistinguishable from the gravel surface of Chinese runways. One of our Douglas C-47 transports flew a whole week on only one rubber tire, a one-inch-diameter Manila rope wrapped tightly around the rim of the other wheel.

I spent the most exciting time of the war in Kweilin during spring and summer; it was thrilling every single morning before dawn to watch crew chiefs locate their planes in the dark (the P-40's were dispersed at night along the edges of the large field). As the engines were warmed up and checked out at full power, bright exhaust flames streaked aft from each plane's twelve exhaust stacks. A terrific engine and propeller roar broke the stillness all around the airstrip. This thrill can perhaps only be understood by those who love powerful aircraft engines. Minutes later, the sky began to lighten and outline the characteristic conical mountain shapes of Kweilin. Crew chiefs taxied their P-40's in front of the pilots' "Ready Tent." When the sun came over the mountains, an open-bed truck brought baskets of scrambled egg sandwiches and coffee from the barracks. . . .

And at night? Of the many air bases we had in China, large and small ones, only Kweilin was repeatedly bombed at night. The reason was simple. Navigational aids were few in the Orient. But a railroad track led from Canton (a major Japanese bomber base close by) to Kweilin. During clear moonlit nights, the two shiny ribbons of track served as perfect guidance for enemy navigators. We usually had two or three P-40's in the air waiting for the Japanese, after receiving alerts from the Chinese ground warning net: "unidentified planes overhead!" Japanese night bomber pilots had a built-in disadvantage which cost quite a few of them their lives: Exhaust stacks of aircraft engines were not shielded aft and emitted bright yellow-bluish flames, making them ideal targets for our planes following them in the dark. One night, Squadron Commander Major Johnny Alison shot down two bombers in five minutes.

By mid-1943, dropping leaflets became a new type of warfare. A plane of the newly formed Chinese-American Composite Wing dropped a message to the Chinese in Japanese-occupied Hong Kong:

> Greetings from your Friends. This leaflet is dropped by aircraft of the United Nations. Every day now more and more planes will fly over Japanese-occupied countries, wreaking vengeance on the Japanese invaders. We are getting stronger every day. Already the Japs are retreating in New Guinea, the Solomons and the South Seas. We are destroying their planes and sinking their ships. Five of their aircraft carriers are at the bottom of the sea. The Allied Air Force in China gets more deadly every day. The United Nations will never forget the sufferings you have endured, the rape of your womenfolk, the loss of your homes, the confiscation of your goods, and the humiliation forced on you by Japanese beasts. But keep your spirits up! Side by side with your brothers of the Chinese Army, we shall come back to Hong Kong.

(Note the reference to the "United Nations." These leaflets were dropped two years before the U.N. Charter was signed in San Francisco.) On the Double Tenth, an annual Chinese holiday celebrated by both Nationalists and Communists of Free China (the Republic of China was born under the leadership of Sun Yat-sen on the tenth day of the tenth month in 1911), a Nationalist training plane dropped leaflets on us, which ended with "We have not forgotten to present you something to brace you up. We hope you share with us the national bliss. Yours respectful."

Not all was "war is hell" in 1943. Soldiers have always looked for an opportunity for relaxation in the nearest town. Our men never failed to locate willing damsels who cooperated, but it was expensive entertainment. Precious gasoline was used to drive to town; tires were cut by jealous Chinese men; drunken soldiers were robbed, and the V.D. rate of Americans in China grew by leaps and bounds. General Chennault had to find a solution to an age-old problem, caused by an age-old profession. Why not ban trips to town completely and instead bring the mountain to Mohammed? Despite the reportedly outspoken unhappiness of Mme. Chiang Kai-shek, who felt the dignity of Chinese womanhood insulted, a wooden two-story building was constructed at the entrance to Kweilin Airfield. On the ground floor was a dance hall with bar and the proverbial piano player "who did not know what was going on upstairs." On the second floor were little bedrooms. There was a medical station, with a mandatory stop for anyone leaving the building, whether he was a saint or a lover. Officers were free to visit Broadway Inn on even-numbered days, enlisted personnel on alternate days—not unlike the odd-even gas rationing of the seventies! But never on Sunday . . .

At that time excellent Japanese and Indo-Chinese beer became somehow available in Kweilin. Nobody dared to ask questions about where it came from or how it penetrated the Japanese front lines; we left well enough alone. There they stood before us: bottles with "Made in Shanghai" labels; others with "Made in Hanoi." It didn't take ingenious GI's very long to find a way to chill the beer without refrigerators. An opening was cut into the top of an auxiliary belly tank; during a "beer flight," no fuel was carried in this special aerial refrigerator loaded with beer bottles. When the plane —exceeding an altitude of 15,000 feet—returned after ten minutes of exposure to −30 degrees F. to −60 degrees F., the beer stored in the belly tank was refreshingly cold.

Nineteen forty-three was coming to an end. The annual rainy season in South China had forced us to move the 76th Fighter Squadron north to Hengyang, in central China. Without prior notice, two days before Thanksgiving, just as it was getting dark, a part of my crew and I were flown secretly to Suichwan, the most southeasterly airfield in China for American planes. Over the past year, Chinese workers had been able to construct a temporary airstrip two hundred kilometers behind the Japanese-occupied railroad line from northern Changsha to southern Canton. Hundreds of fifty-gallon fuel barrels camouflaged with rice straw had been moved through enemy lines on two-wheeled wooden carts pulled by water buffaloes. The essential secrecy was obviously highly successful. Neither we nor the enemy knew what General Chennault had planned. Next evening, a squadron (fifteen planes) of old Curtiss P-40's landed on the dirt-and-gravel runway outlined by twenty hand-held flashlights. Thirty minutes later, twenty of America's newest, faster and longer-range North American P-51 "Mustang" and Lockheed P-38 fighters showed up; the P-51's had arrived only the previous day in China, and came straight from the North African campaign. Within minutes, a squadron of medium-sized Mitchell B-25 bombers with large 75-millimeter cannons protruding from their noses landed at Suichwan. Pilots brought with them yellow lifesaving vests. What was all this about? Neither the pilots nor we knew. Colonel "Tex" Hill, a former original Flying Tiger and a fighter ace, was the commander.

An unexpected problem arose: The P-51, much more streamlined than our old P-40 warhorse, had its engine coolant and oil radiators located at the lowest point of the fuselage, aft of the cockpit and not behind the propeller high off the runway like the P-40. No doubt this was aerodynamically a splendid idea, but a disaster in the real world of Chinese runways, made of rocks and gravel instead of asphalt or cement. As a consequence, and very similarly as in the case of my Zero's accident in Kweilin eighteen months earlier, the air blast of the propeller while on the ground threw tiny pebbles aft with tremendous force, puncturing several P-51 honeycomb oil and engine coolant radiators. The result was critical leakage of essential cooling fluids. Nearly all of the newly arrived P-51's were in no condition to fly anywhere the next day unless we replaced their damaged radiators—an impossibility. Something had to be done right then and there.

I had an idea that today sounds crazy. We sent four of our American-Chinese mechanics to a nearby village (the residents of occupied China were loyal to Americans) to purchase from them four old bicycles. We removed the wheels, took out the spokes, cut threads at each end of them and inserted the spokes into the leaking honeycomb passages of the radiators. The leaky passages were sealed at each end with two rubber washers cut from the bicycle tires' inner tubes and tightened with nuts. This fix worked!

Next morning, very early on Thanksgiving Day 1943, our P-40's left our temporary airfield, heading north to fool anyone who might possibly report to the Japanese intelligence the goings-on at Suichwan. Thirty minutes later, all P-51's, P-38's and B-25's took off in tight formation in a southeasterly direction, toward the Formosa Strait. Only then were we told about the mission and understood why the pilots wore life vests. The flight was a long one. It was imperative that the most direct route be flown to conserve every bit of fuel—and to be able to return to Suichwan. We could barely stand the suspense as hour after hour went by. Total radio silence made it impossible to follow the progress of the surprise raid. Would it work? What would our planes find at their "Target Formosa" (today Taiwan)? For years American reconnaissance had shown to General Chennault photos of masses of fighters and bombers staged on the airfield there. That island had been occupied by the Japanese for fifty years. These Japanese knew that Formosa was beyond the range of P-40's, which therefore could not fly top-cover for American attack bombers. That's why the Japanese used "safe" Formosa, with its excellent runways and complete aircraft service facilities, as a staging area for planes to be ferried from Japan to islands in the South Pacific.

Our fighters and bombers were carefully scheduled so as not to interfere with each other: They had to arrive over the target in a tight sequence, at precisely the right moment. Shortage of fuel did not permit any mixup. True to the preflight briefing, our pilots found Japanese fighters and bombers lined up, wing tip to wing tip, in double rows, on both sides of the runways. Our P-51's, P-38's and B-25's had skimmed so low over Chinese rice paddies and the Formosa Strait that the Japanese radar defense system was never even alerted. The surprise was 100 percent successful! One lone Zero got off the ground at the last moment but was shot out of the sky before it even had a chance to pull up its wheels.

Three and a half hours after takeoff from Suichwan, more and more specks showed up on the horizon. We excitedly began to count our planes as they touched down on our temporary base. *All* had returned safely! The P-40's that had left earlier for a fake attack to the north had returned three hours before, had been refueled, and were now back in the air, circling and protecting Suichwan's runway. We quickly readied the returned fighters and bombers in case the Japanese sent planes to retaliate, catching ours on the ground with our pants down.

That precaution turned out to be unnecessary. The Japanese must have been so shaken by the surprise raid from the Chinese mainland that they had no thought of retaliation. The only error made by the Americans was an incorrect guess of the wind direction over Formosa, which caused the thick black smoke of burning planes to blanket part of the airfield and thus saved a few of the Japanese planes. We had, according to photos taken by our high-altitude reconnaissance Lockheed P-38 two hours following the raid, totally destroyed sixty-three Japanese aircraft and seriously damaged many more. Not one American plane had a bullet hole in it. When we returned to our home base on our side of the front late that Thanksgiving Day 1943, we received a message from a very happy General Chennault: "THANK YOU ALL. THANKSGIVING TURKEYS WILL BE SERVED A DAY LATE." He had ordered a C-47 transport loaded with turkeys to be flown from India to our Hengyang base in China; each man who had been at Suichwan feasted on turkey as promised. We engineering folks said a little prayer of thanks that the bicycles we had taken apart permitted all P-51's to participate in this successful raid and to return safely.

I was looking for something else to do during the bad weather season and was transferred to the 322nd Troop Carrier squadron as aerial engineer. Practically daily, we visited all bases we had in China, flying in all kinds of weather. One time, near the central city of Hengyang, which was close to the Japanese lines, we dodged three of their Zero fighter planes by escaping at tree- and rice-paddy-level altitudes that made our unarmed but camouflaged C-47 nearly invisible. Flying was fun but lasted only a few months; an urgent requirement for my services showed up some place else.

Over the years I had picked up from Chinese soldiers a fair amount of their "GI language" (quite different from the classic Mandarin Chinese taught at U.S. language schools). I was able to converse with them adequately to get along. Because of that, I was asked to transfer to the 5329th AGFRTS (Air Ground Forces Resources & Technical Staff), a branch of the OSS (Office of Strategic Services), the wartime CIA. For eleven months I was a member of various two-men field teams whose task it was to report on the goings-on of not only the Japanese enemy but also of our Chinese ally. (One of my team- and tent-mates was Lieutenant John Birch, the mild-mannered son of a missionary who was born in China and spoke its language fluently. The reactionary John Birch Society of the sixties was named after that late officer, without any obvious reason. Lieutenant Birch would have shuddered at the thought of being a part of that society.)

We made week- to month-long trips into the vicinity of or behind enemy lines. We traveled by river junk, Jeep, or on foot. On one mission, a Navy captain and I wore coolie hats and carried bamboo poles over our shoulders with straw baskets swinging at each end. We walked for days on narrow, slippery footpaths between rice paddies, ate whatever we could find, and slept in farmhouses usually protected by a circle of Chinese guerrillas dressed in black suits and armed with German Mauser pistols. One of our tasks was to direct air strikes against the Japanese via radio. A communication set had been installed in our Jeep, which we hid in bamboo growth overlooking the enemy lines. Electricity for the radio transmitter came from a hand-cranked generator either of us two agents had to crank wildly, a job more wearing than the spy mission itself! The enemy knew very well that someone was directing our American planes to their hiding places but was unable to figure out from where the guidance came since we constantly moved our Jeep. Boats and ferries loaded with Japanese troops, light tanks and cavalry were our favorite bombing targets. We suggested the number of our aircraft to use and what type of bomb to drop, and gave precise directions when to leave the home base to catch the Japanese in the midst of their river crossings. We talked in code while the planes were still at home base and en route and we had the time it took to consult the agreed-to codes; but we switched to plain English when communicating with the bombardiers flying overhead, often as low as 1,000 feet. By following our directions from the ground, they knew precisely when and where to drop antipersonnel or explosive bombs even if our aircrews could not see the targets from above.

Ground defense of air bases had been assigned to the Chinese Army per the agreement of General Chennault with Generalissimo Chiang Kai-shek. Most soldiers were farmers conscripted from their rice paddies, which they had left reluctantly. (I saw some draftees, their arms tied tightly behind their backs, a Manila rope knotted from one neck to the next neck, being led away for induction and training to become defenders of their country.) Chinese forces were brave but most were poorly led; troops had few weapons and even less ammunition. They were not permitted to fire the few individually numbered artillery shells unless granted permission by their commanding officer. Once, I stared in disbelief at Japanese troops preparing to cross the Siang River on barges near Lingling, one of our valuable air bases evacuated half a day earlier. The Chinese army had at the ready two beautifully positioned and well-camouflaged artillery pieces that were set up to fire point-blank 500 yards across the Siang and which would cause death and destruction. At the critical moment when the Japanese launched their ferries and began their river crossing, a Chinese soldier was desperately trying to find the officer in charge whose okay was required before the Chinese troops could shoot. The officer reportedly had left in a sedan with a young lady, destination unknown. He could not be found. Ergo: The Chinese artillery withdrew from the river without having fired a single round; the Japanese crossed unimpeded and a few hours later took possession of the Lingling air base two miles from the river. This situation was not unique. During summer and autumn of 1944, the Japanese were able to cut off huge sections of Chinese territory, occupying our American air bases and taking quantities of supplies which we had no time to destroy. (I saved a couple of cans of strawberry jam!)

When we two agents returned to Kweilin from that assignment, we found our main base at Kweilin in the process of being scorched by Americans left behind to destroy the supply shacks, barracks and both runways. This took place only one week after we had lost Lingling and two weeks after pulling out of Hengyang. It was obvious that the Japanese were determined to eliminate the American Air Force in China, which had been a constant thorn in their side.

Danger signals of collapsing morale were apparent during the Chinese war briefings I attended in summer and autumn of 1944.

Reports by Chinese junior officers to their field commanders did not always jibe with facts. For example, rail lines were once reported as "torn up from *A* to *B* and dumped into the nearby river." If true, that would indeed slow down the Japanese advance and give our American Air Force some extra time before evacuating bases, or to utilize them for a few more raids. But it wasn't true. The Chinese commander, of course, did not know this and may well have made a perfectly correct strategic decision based on the facts given him by his officers, but those facts were inaccurate or outright false. (A most important lesson for my own future career: "Get the facts!")

In mid-October 1944, a radio message from General Chennault's headquarters in Kunming ordered me to report to him as soon as possible. The monsoon season had just begun and rendered flying in eastern China impossible. There was no longer any air support available for the hard-pressed Chinese ground troops. The OSS captain with whom I was traveling to Kunming requisitioned one of the scarce evacuation Jeeps. He and I slithered along 550 miles of muddy, unpaved roads, averaging fifteen miles per hour. We carried along enough gas in jerry cans to make the trip nonstop. Driving day, night and day, taking turns at the wheel, eating cold C rations, we reached the 14th Air Force headquarters hungry, unshaven and tired, looking like characters in Bill Mauldin's famous cartoons.

I was ushered immediately past the General's aides into the same simple mud-floored office he had occupied as a retired U.S. Army captain when I met him the first time the summer of 1940. Now Commanding General of the 14th Air Force, Claire Lee Chennault questioned me in detail about my recent experiences with the Chinese forces and asked what I thought about their ability to hold their ground.

"Negative!"

After an exchange of questions and answers, Chennault rang his desk bell for his aide. "Colonel Strickland," he said, and pointed at me, "this is Master Sergeant Neumann; he has just returned from the east. I want him to go to Washington right away and report to General Donovan, head of the OSS. Have his travel stencil cut, please."

"General, I know about Sergeant Neumann. He certainly deserves to go to America.'' The colonel hesitated, then in a barely audible tone said, ''But Sergeant Neumann's trip to the States would be illegal, sir . . .''

''And what in hell is wrong with getting him into the States *illegally*?'' thundered the General at his aide.

10

First Visit to America

The sergeant behind the wooden counter at the Kunming Air Transport Command shack didn't quite believe what he had read. He had never seen such brief, mysterious travel orders—signed by the Commanding General himself! In less than a minute I was assigned to a Curtiss C-46 transport shuttling between China and Assam. The plane's right engine was already running as I climbed aboard. A toothbrush, shaving kit, green washcloth, green towel, green underwear, green handkerchiefs and my camera were all that I carried with me on my first trip to America. My buddies in China —whether enlisted men or officers—knew that I'd never been to the States or had any acquaintance there. Over the past few years the fellows had given me addresses of their parents, wives, sweethearts, brothers and sisters, to give them a ring or look them up should I ever get to the States. "Tell them you saw me, and you'll get the biggest steak you've ever seen!" they promised. My little black book was neatly organized by states and cities, with names of the soldiers' relatives, their addresses and phone numbers. I stuck this valuable entertainment pass into the left breast pocket of my flying suit.

The Curtiss "Commando" C-46 I jumped on was a cargo plane carrying tin and wolfram to Burma, where it was unloaded. Its cargo was then sent on to the USA. It was not set up to transport passengers. On its return trip from Assam it would bring to China aviation gas, bombs and ammunition, the same matériel that I had hauled to China in the Renault truck convoy over the original Burma Road in 1940, four years earlier. Cargo air transport was vital because the new Ledo Road being built by American Army engineers farther to the north and out of reach of the Japanese, replacing the cut-off Burma Road, was not yet ready to be put into service. The flight crew warned me that I would be breathing the minimum of oxygen needed to stay alive in the unpressurized cabin at 21,000 feet, for two and a half hours, and alerted me to be prepared to pass out during the flight. They bundled me into a fur-lined flying suit and made me lie down on top of the metal bullion. After takeoff I sat up to take a last look at Kunming, its lovely Dianchi lake, the surrounding mountains and rice paddies—then passed out. When I came to, humid jungle air was streaming into the plane; a turbaned Indian bearer had opened the wide cargo door and was staring at me. We had landed at Dinjan in Assam (northeast India) after a three-hour flight over the Hump.

Sixty minutes later, I left in another Curtiss C-46 transport for Casablanca, on the North African coast bordering the Atlantic Ocean, via Agra, Karachi, Cairo and Tunis. It was warm and comfortable flying a few thousand feet above the flat, brown and dry Indian terrain. Approaching Agra really low, we admired the exquisite white marble Taj Mahal with its four slender minarets, built in the seventeenth century. Agra was the biggest U.S. air base in Asia. A large number of new fighters, transports and bombers were lined up along both sides of the runway, to be distributed to U.S. forces in Burma, China and the Philippines. Before reaching Agra our plane had developed a propeller problem, whose repair was expected to take only an hour. Across the taxi strip was a large American Red Cross tent offering free doughnuts and Coca-Cola. Sitting inside on empty ammunition boxes, GI's were taking each other's money at poker and blackjack. Two soldiers were playing chess at the rear of the tent. I kibitzed for a while, then challenged the winner. He turned out to be an ex-German, on his way to China with the 20th Air Force (whose sole mission it was to bomb Japan). Like all GI's meeting anywhere, we exchanged the routine "where from?" and "where to?" I told him of China, and he briefly described New York and Washington. When I asked if American girls played chess, he gave me an odd look . . . then remembered that he knew one who did, an attorney with the Department of Justice. He confirmed what the boys in China had told me: There were so many single girls in D.C. that "a man had to beat them off with a club." The girl-to-boy ratio was something like 10 to 1. (What could I do for the remaining nine? I wondered. . . .) I jotted the name and address of that chess-playing girl onto a slip of paper and stuck it into the zippered side pocket of my flying suit. We had not quite finished our second chess game when the public address speaker called me to return to my C-46.

The next unexpected delay occurred at Cairo. Because of scheduled plane maintenance we had to stay there six hours. I took an Army bus to town and was surprised at the modern buildings next to old quarters and minarets en route. Ancient boats with triangular cotton sails moved slowly across the Nile. In 1944 a streetcar line went from downtown Cairo to within a few hundred feet of the Pyramids and the Sphinx, whose chin was supported by sandbags against bomb concussions. Egyptian youngsters roamed the vicinity of the streetcar terminal singling out GI's. Two boys ran by me and spilled white paint on my shoes. Annoyed, I was just going to run after them when two of their associates appeared with material to clean my shoes. It was a demonstration of teamwork—for some baksheesh, of course!

At the largest of several pyramids, I joined thirty GI's admiring it in awe: How were the workers able to move those huge blocks of stone, one on top of the other, thousands of year ago? We GI's then wanted to see a "real" mummy; holding on to each other, we were led by a guide up a dark, narrow, and upward-sloping tunnel into the pyramid's geometrical center. Deep inside, the steep passage leveled out. It was pitch-dark and impossible to see anything. Our guide offered to light up the chamber for a few more piasters (the odd shapes of Egyptian coins could be felt easily in the dark). He got his ransom, and lit a red magnesium flare that burned so bright it blinded us from the moment it was lit until half a minute after it died out—and we still hadn't seen anything! Then for a few additional piasters (worth only pennies) he switched on the bare electric lightbulb installed some years ago. A stone sarcophagus without a cover stood in the middle of the damp cube-shaped

4,500-year-old grave. We rushed over to it. It was empty. The mummy of the Pharaoh who had been buried in this pyramid had been moved to the British Museum in London decades ago, the guide explained. We teetered down the narrow path and emerged, squinting in the brilliant desert sunlight.

A Bedouin pestered me to pose for a photograph on his camel, stretched out on the sand. I gave him a few coins and climbed onto the saddle. I never before had paid any attention to how a camel gets up; I was unaware that it raises its hind legs first, lifting its rear high into the air, then follows up with its front legs. As the lumbering creature rose camel-fashion, I was thrown forward and slid down its neck all the way to the ears, barely managing to hang on. Later, during the flight segment from Cairo to Casablanca, when I looked for the little black address book I found that it was gone! I knew that it was still in my open breast pocket when I climbed onto the camel; it probably had fallen out when I hung, nearly upside down, on the camel's neck. This was a terrible blow for a tourist who knew no one in America!

The C-46 flew 1,500 feet high along the North African coast to Tunis, where I had landed in an Air France plane five years earlier, going in the opposite direction. Below us were clearly visible masses of wrecked tanks outside the fortress of Tobruk, where the German and British desert troops fought two bitter battles in 1942. From Casablanca I flew across the Atlantic in a four-engine Douglas C-54 VIP luxurious transport via the Azores and Bermuda. An hour before arriving in the States, a crew member distributed customs/immigration forms with a dozen questions. Typical queries: Home address in the States? Any foreign currency in your possession? Anything to declare? At "When did you leave the States last?" I wrote "NEVER."

Hurrah, America, here I am! On October 26, 1944, at 0630 the weekly courier plane from India landed at New York's La Guardia, under a cloudy sky. The highest-ranking officers got off the plane first. As the only noncommissioned officer aboard, I was the last man off the C-54. A two-man welcoming committee from the FBI was awaiting me. Over coffee and doughnuts in a nearby shack, they asked if it were true that I had NEVER left the States before. Yes, it was. And that I had no address in the States? Yes, it was. (The plane's crew must have radioed ahead about the odd GI aboard.) Where was my passport? In Hong Kong, with the British—or maybe with the Japanese? What was I going to do here in the States? "That's secret!" They looked at my travel orders, shook their heads, quietly talked to each other; then one of them disappeared to phone Washington from the adjoining room. General Donovan's OSS headquarters obviously gave me a clean bill of health and told the FBI agents that I was expected. Did I get the royal treatment after that! A staff car was called, a reservation made for me at the Sutton Hotel in New York City; a female in a dark-green U.S. Army uniform (my first encounter with a WAC) chauffeured me to the downtown hotel, which had been taken over by the military to provide temporary quarters for its transient personnel.

It was shortly before 10 A.M. when I registered. While walking down the hall to my room on the assigned floor, I passed an open door leading to a white-tiled bathroom. My room was not much farther down the corridor; small but adequately furnished, it had a window and two side doors. I opened one: a closet. I did not bother to open the other door, but undressed quickly, wrapped my green towel around my waist and hurried back toward the white bathroom, whose door was still open. I could hardly wait! My first bath—in a real bathtub . . . in how many years? Even in Calcutta's base hospital a year earlier one could take showers only. It was heavenly to relax in the hot tub, as I remembered my last shave in a rice paddy, with brown, muddy water in my helmet and leeches clinging to my hairy legs—only a few days ago! I was still soaking in the tub when someone pounded at the bathroom door. "What are you doing in *my* bathroom?" Perplexed, I hurried out of the tub, dried myself, and had some explaining to do to an impatient lieutenant colonel! It turned out that the second side door in my room led to my own bath! In prewar Europe one was lucky if there was one bathroom for each hotel floor.

At a phone booth in the Hotel Sutton, I looked up the number of a New York radio actress, Ann M., the sister of a major in my squadron. I had promised him I would give her a call as soon as I got to New York. Fortunately, I recalled her name without my black book. She was at home. "Oh, yes, Sergeant, I heard about you from my brother—I'm looking forward to meeting you. How about six o'clock in the lobby of the St. Regis, under the mirror?" How would I recognize her? "Don't worry," she said "I'll find you, and you'll recognize me." I didn't tell her that I did not know what a lobby was.

There was a message for me at the desk from a colonel of the Army Public Relations office. The officer would be delighted to meet someone newly arrived from the war in China, it said, and he invited me for lunch. When the colonel saw me in my flying suit and learned that I had no other uniform with me, he arranged for me to get two sets of khaki uniforms with master sergeant stripes and wings sewn on, by 3 P.M. that afternoon. Next, he defined the word lobby for me, not contained in my tiny, very much abbreviated English-German pocket dictionary. The reason for all his attention: He wanted me to appear at a press interview in his P.R. conference room not far from the Sutton next morning at ten. The colonel would sit next to me, he said, and kick my leg when I was supposed to say "I *really* don't know . . ." (whether or not I knew did not matter). Otherwise I was to answer all questions truthfully. He also suggested that I go to my 6 P.M. date at the St. Regis via subway if I wanted to see New York at its best. At this point in time, on my first day in the United States, I was already overwhelmed: the skyscrapers; my own bath; the mass of white civilians, taxis, buses with windows, two new uniforms! Here was a New World for me, literally!

After being fitted with new uniforms, my Flying Tiger pin at the right breast pocket, the USAF wings on its left, I was ready for anything—and it didn't take long to come. To avoid confusion I planned ahead: I separated nickels and dimes in my right and left trouser pockets. (Why is a five-cent piece larger than a ten-cent one?) In the nearest subway station, at the beginning of the rush hour, I was hurried along by the mass of commuters and shoved into a row of turnstiles. I dropped a coin into the slot of the turnstile but it did not budge. Another coin, again nothing happened. Then another one. The man behind tapped me on the shoulder. "Soldier, you are putting the coins in the wrong turnstile!" I had been paying the fares for the people next to me, who were happily pushing their way through. At 5:45 P.M I reconnoitered the area below the large mirror in the St. Regis lobby where I was to meet Ann M. Except in movies I had never before seen such a fancy place, such chandeliers and elegant furniture! The lobby was getting crowded with matronly ladies and older men.

Through the crowd emerged a fur-draped stunning platinum doll, heading straight for me. Yep, this was she, Ann M.! My heart skipped a few beats as she took me by the hand and towed me to the adjoining hotel bar. I had never seen an American bar, let alone sat on one of its high stools. "What will it be, soldier?" the bartender asked. I had not the slightest idea what to order; except for beer or Chinese rice wine, I had never drunk any alcohol. When I hesitated, my beautiful companion asked sympathetically, "Sergeant, do you prefer a hard or a soft drink?" Her question was totally incomprehensible to me. Hard? Soft? To me all drinks were liquid! For herself, she ordered something "double"; to make a quick decision, I asked for a Coke. . . .

We were talking about China and her brother when Miss M. pointed toward an elderly gentleman entering the bar from the lobby. "Here comes Sugar Daddy!" (Who?) After a brief introduction I was beginning to relive with him the flying days of Captain Eddie Rickenbacker and his 94th Squadron in France, of which this sugar daddy had been a member in 1918. The gentleman, president of a company whose name I had heard in China, invited Ann M. and me to a delicious dinner in the sumptuous dining room upstairs. I hadn't selected something to eat from a printed menu for years! A glass-covered dessert cart was rolled to our table. After rice and sponge cake for the past three years, Viennese pastry and chocolate pie were quite a treat. When I said good night to both of them, Miss M. asked me to call her the next time I'd be in New York—but I chickened out! I never saw her again.

At nine next morning, the colonel picked me up for the press conference. There were twenty-two reporters from the *New York Times, Herald Tribune, Post, Daily News,* German-American newspapers in English and German, and many more. An hour-long interview, questions and answers primarily related to my being in the American Army and why we had been losing so many air bases in China recently. The colonel had to kick my knee only twice, when questions relating to Mme. Chiang and the Generalissimo came up. Photos were taken after the interview.

I bought one each of next morning's newspapers. When I read what the reporters had to say about our conference, I nearly died! Incredibly different stories came from one and the same interview; many articles were not only untrue, inaccurate and exaggerated, but outright sensational. I was terribly embarrassed. Of all the newspapers, only two were accurate. In the next issue of *Life,* the back cover of its overseas edition sported a full-page photo of my head and shoulders, with the caption "G.I. in a Jam," reporting correctly that I had just arrived from overseas as a U.S. Air Force master sergeant and was an ex-Flying Tiger, but could not get a commission because I first had to obtain U.S. citizenship. A United Press reporter invited me for a day-long car ride next day to see other parts of New York City, from Harlem to Wall Street and to lunch at the Waldorf-Astoria. Since the colonel had told me earlier that my trip to Washington would have to be delayed by one day (because no Pullman reservation could be obtained until then), I was glad to accept the reporter's invitation to see more of New York.

Once arrived in Washington, I had an immediate appointment with famed General Wild Bill Donovan, winner of the Congressional Medal of Honor in World War I. He had received a message from General Chennault, outlining my background and requesting him to do whatever he could to get me a "direct field commission" since I handled classified material. The friendly and informal OSS general, a lawyer in civilian life, welcomed me at the door to his office. I described to him the current state of the Chinese forces as I had observed them. He listened attentively, then summoned his aide (a Navy captain) and ordered him to see that I was well taken care of during my first visit to this country, and that I did receive the commission as an officer. For an extra welcome bonus he directed that I get a blank, preauthorized pad of ten three-day passes, which I myself was to fill out wherever I might be at those dates. I thanked General Donovan and promised to see him again in sixty days. I remained with his aide, who called a government lawyer and requested him to research what it would take to do what the OSS general wanted.

"Where to next?" asked the aide.

Thinking of a friend of the Calcutta Red Cross girl, Virginia, I replied, "To Chicago." The captain arranged for a train reservation leaving Washington in a few days. That same afternoon I looked up the address near Rock Creek Park of the chess-playing girl, given to me by the soldier I met playing chess in India. Because I had written her address on a separate slip of paper, it was not lost as was my black book at the Pyramids in Cairo. No one answered her doorbell. I waited for a while, sitting on a hallway step, then walked to a nearby corner drugstore—the very first American-style drugstore I had ever been in, i.e., in which one could buy just about everything besides drugs. A young waitress concocted for me a giant ice-cream sundae: black, white, pink and yellow, topped with marshmallow, chocolate sauce and nuts . . . *the works*! Her boss came over to see what was going on—then refused to let me pay.

Before returning to downtown Washington I went one more time to see if the chess-playing lawyer had come home while I was stuffing myself. She had. Clarice's pretty little face with her shiny, bright and attractive eyes peered out from behind the chain holding her door slightly ajar. From the hall I passed on greetings from her friend of the 20th Air Force whom I had met in India five days before; only then did she let me into her apartment. Clarice had just returned from donating blood for the Red Cross and she accepted my dinner invitation for that evening. With so many "surplus" girls in Washington, all out "to please the soldier in uniform" (so they said, anyhow), it was easy to have a date with a different girl every evening. Clarice even urged me to use her 1940 DeSoto coupe for a dinner date she would set up for me with a young office friend of hers.

A few days later, I set off for Chicago by train. Around midnight military police checked soldiers' passes in the club car. I filled out one of the three-day pass blanks of my presigned pad right on the spot, causing quite a stir amongst the MP's. At Chicago I found a room in a southside hotel for $1.50 a night, then phoned the Chicago businessman. I had no idea who this man was except that he was a friend of Virginia's. First one, then another secretary answered the phone; finally the man himself, Mr. Duane T. Molthop. "Oh, yes, young man, I heard about you from Virginia. See you at noon tomorrow under the clock in the Union League Club." I, of course, had no idea what a swank place the Union League Club was. A big, six-foot-tall, good-looking man with white hair and blue eyes, perhaps forty-five to fifty years old, arrived in a long black limousine, in spite of the gasoline rationing. He patted me on the back, shook my hand firmly and at once made me feel at ease. We had a fine meeting and a great lunch. "Bring the biggest steak you have," he told the headwaiter. "This soldier has been eating buffalo meat for the last three years." Mr. Molthop invited me to another lunch, another dinner, another lunch, and another dinner: Union League Club, Chicago Athletic Club, Pump Room of the Ambassador West, Ambassador East—and then to stay at his home in LaGrange, outside Chicago. Molthop was (and still is) the president of the Vierling Steel Works. His three sons were overseas with the U.S. Navy. To one of the dinners he had invited good-looking Betty Willkie, about twenty. Betty was the niece of the U.S. presidential candidate Wendell Willkie. There followed more invitations to the homes of some of America's leading industrialists in exclusive LaGrange and my first dinner in a country club. Before I returned to Washington, Mr. Molthop presented me with a dozen rolls of 120 camera film, which was difficult to get.

Back at the OSS barracks on Twenty-third Street in Washington, I found two tickets for the annual Army–Navy football benefit game in Baltimore, with seats on the fifty-yard line. The tickets had come from my thoughtful friend Molthop in Chicago: "For you, and whoever your favorite lady may be. . . ." This was the first football game I had ever seen. Totally ignorant of the rules, I could not understand why all those discussions took place on the field every few minutes, why the "clock stopped," why the teams were falling over each other—"football" was hard to follow for one who was brought up playing the straightforward, orderly game of soccer!

I had another challenge every morning, when I ate breakfast in a drugstore: I ordered Rice Crispies. Each and every time the waitress asked, "What?" I switched my order to Corn Flakes. My *r* could evidently not be understood.

I met total strangers who, when they learned that I had no home in the States, invited me to stay with them. Americans were truly wonderful hosts to this enemy alien. I led an unrealistic, whirlwind life: from a cocktail party at the apartment of the editor of *Seventeen,* to a skiing weekend in the Catskills. . . . During Christmas 1944, a naturalized ex-German general whom I had met at one of these parties and who was going away over the holidays even gave me the key to his apartment on Central Park in New York. Over his radio I heard the news that the last serious German counteroffensive of World War II, at Bastogne in Belgium, had been halted, and that—on the other front, in the east—the Russian steamroller had broken through the German lines at many areas in Poland, and was approaching my hometown Frankfurt/Oder, the last bastion before Berlin.

During the next few months I saw more and more of the chess-playing girl. We played Ping-Pong, rode horses in Washington's Rock Creek Park; she loved dogs as much as I did and had a great sense of humor. Her 1940 DeSoto club coupe wasn't being used much because of the gasoline shortage. I became a valuable asset to her with my potential for thirty gallons of gasoline: Each GI was entitled to one gallon of gas for each day of his furlough whether he owned a car or not.

One weekend in January 1945, Clarice and I met in New York's Central Park, visited the zoo, ate hot dogs (which, I realized, *she* was not used to) and stood in line for an hour at the New York Opera, trying to get tickets for an evening performance. There were about twenty more people ahead of us when the sign SOLD OUT was lowered over the cashier's window. A gentleman behind us asked if I had seen action overseas. Hearing that I had, he insisted that we accept his own two orchestra seat tickets (which he was going to exchange for some other date anyhow). He absolutely refused to accept any money from us. Another time, when I went sightseeing alone around the Wall Street area, an old Irish cop on the beat invited me into the Stock Exchange and tried his best to explain to me what all the yelling and hand-waving was about.

Thanks to introductions by General Chennault and Virginia, I met celebrities of the nonindustrial/military circles, such as Supreme Court Justice Frank Murphy, a former U.S. high commissioner of the Philippines. He invited me to lunch in his chambers and proudly showed me the scrapbook of his National Guard service with an armored division in which he was a colonel. Before the afternoon session of the Supreme Court began, he introduced me to the other eight Justices and invited me to attend one of the Court's sessions. I accepted and heard the case that afternoon, *United Fruit* v. *Southern Railroad,* regarding bananas damaged in transit during train shunting operations, a case removed from the war and China. When Clarice asked me that evening "What did you do today?" and I casually told her about my having had lunch with Justice Murphy, she was certain that I was kidding! . . . Through another wartime acquaintance, I met Miss Toinette Bachelder,

President Roosevelt's personal secretary, who invited me to join some of the President's staff for dinner at her elegant Georgetown apartment.

One evening, Clarice and I decided to splurge on a festive dinner in the dining room of the old-fashioned Raleigh Hotel. The waiter who had taken our order pointed to a table across the dance floor and said, "That gentleman over there wants to have the honor of inviting you and your date as his guests." No question about it: Americans were truly wonderful to their soldiers and sailors—especially to those who had returned from overseas, easily identifiable by colorful campaign ribbons worn on the uniform.

Six weeks after my first visit, I went again to see General Donovan. He gave me the bad news that in spite of efforts by the best brains of the Defense Department, I could not become an officer as requested by General Chennault because I was legally still an enemy alien rather than a cobelligerent or a neutral. As a matter of fact, the State Department very quietly pointed out to the OSS that without a passport and without a visitor's visa, I was in the States illegally. There was nothing that could be done now, General Donovan said—yet he would continue to try to resolve this absurd impasse ". . . . even if it takes an Act of Congress!" This statement I did not take seriously since our supply sergeant in China used the same expression if you asked him for anything like an extra pair of socks. Disappointed but realistic, I wrote a letter to General Chennault asking him to recall me to China, although General Donovan had pointed out that I had accumulated more than enough "overseas service points" to remain in the States. I didn't feel right in wallowing in America's luxury with my only daily concern being which girl to invite for a bicycle ride or whom to take out for dinner that evening. I told General Donovan that I'd rather be a master sergeant with Chennault in China than a captain in the United States, anyhow!

During the working day I was assigned to the Naval Technical Air Intelligence Center at Anacostia; I gave advice and counsel on Japanese flying boats and bombers, of which we now had captured quite a few. I spent two weeks at the OSS training camp outside Washington at the Congressional Country Club, learning hand-to-hand combat from a colonel of the prewar Shanghai police. A cable arrived from General Chennault in March 1945, ordering me to "return to China by first available transportation" and to report to his headquarters.

It was April 1945 when Clarice drove me to Union Station. The time had arrived for thanks—and good-bye. We had agreed not to look back when she dropped me off at the station and I passed with my duffel bag through the swinging gate of the railroad platform. It had become obvious to me that I was far more fond of Clarice than of any other girl I had met. I never thought that it could really happen to me: Yes, there was a bit of a lump in my throat.

11

U.S. Citizen by Special Act of Congress

The war in Europe was coming to an end. Hitler's boast of a thousand-year Reich lay in shambles. All over the Pacific, too, the war was going well for the Allies—except in China, where the Chinese Army was unable to stop the Japanese. When I arrived in Cairo in April 1945, fourteen hours after leaving the States, word of President Roosevelt's death had spread. A pity, we soldiers felt, that our Commander-in-Chief could not be alive to witness the ultimate Allied victory.

One week later, after I stopped at Agra to tell Virginia, my Red Cross friend (who had been transferred there from Calcutta), about my American visit and to see the Taj Mahal in full moonlight, I was back at Chennault's headquarters in Kunming. I reported to the Old Man on my trip to the USA, transmitted greetings from the General's friends who had given me such a wonderful reception, and thanked him for all he had done for me. With Germany now on the verge of collapse, it was expected that the final action of World War II would take place along the Chinese and Japanese coasts. Millions of Allied soldiers, primarily Americans, British and Australians, were rumored soon to be landing along the southern and eastern parts of the Chinese mainland. I told the General that I would like—for sentimental reasons—to wind up the war with the same outfit with which I had started in 1942. He understood, had me removed from the OSS and assigned to a newly formed 118th Reconnaissance Squadron that had become a part of my old 23rd Fighter Group. This squadron was commanded by Major Marv Lubner, who had been my first engineering officer three years before when he was still a second lieutenant. Now, in 1945, he was an Air Corps ace and on his second tour of duty in China. Except for Major Lubner and his commanding officer, Lieutenant Colonel Ed Rector (an ex-Flying Tiger ace and my squadron commander in 1942), none of my old buddies was left in China.

One day in May 1945, a cable arrived for me from Washington: "CITIZENSHIP BILL INTRODUCED IN CONGRESS TODAY. CONGRATULATIONS. CLARICE." Unbelievably and totally unexpectedly, General Bill Donovan had kept his word that he would try to overcome the conditions which made me ineligible for U.S. citizenship and a commission. He had interested New York Congressman Andrews, of the House Military Affairs Committee, in introducing a bill waiving the requirements for naturalization in my specific case. I never believed that this would happen: an Act of Congress on my behalf! Nor that Clarice would have enough interest in me to follow the matter in the *Congressional Record.*

In view of the Allied progress in the Pacific and the expected reinforcement of American troops from Europe, now that the war against Hitler was over, the Japanese halted advances on all Chinese fronts. Their high command began to pull back forces from China's interior in preparation for the defense of her occupied coast and their own homeland. U.S. airfields taken from the Chinese as late as November 1944 were once again changing hands. Scorched a second time, now by the retreating Japanese, their runways were being rebuilt and craters from buried bombs filled in for use by us.

At four o'clock in the morning of August 7, I was awakened by the loud voices of a couple of soldiers outside my tent: ". . . the bombardier fell off his seat when the bomb he had dropped over Japan from twenty thousand feet exploded!" I called out to the fellows to shut up and let us sleep; but one came to the rolled-up side of my tent and whispered excitedly through the mosquito net, "It's true, Sergeant! We dropped a bomb on Hiroshima and all hell broke loose!" By that time the five other men in my tent were awake, and so were GI's nearby. We speculated that those two fellows outside must have had a little too much rice wine or read too many comic strips about ray guns and Buck Rogers. At breakfast time rumors were beginning to flourish in our tent site adjacent to the fighter strip: Something special *had* happened—that was sure. One of our radio operators picked up news from the Armed Forces radio in San Francisco that the Pope had sent an urgent message to President Truman. Still, none of us believed these fantastic rumors. Three days later came more exciting news: Another "incredible" bomb had been dropped on a second Japanese city, Nagasaki. The whole city reportedly went up in flames. From a single bomb? Unbelievable. Shortly thereafter, a broadcast to all American troops, wherever stationed, was received from our new Commander-in-Chief, President Harry Truman: "THE WAR IS OVER! Japan has surrendered unconditionally!"

Although we shared the relief of the millions of troops around the world who reportedly celebrated by firing guns into the air, our fighter air base in China remained totally silent. Many GI's in the CBI theater of war, particularly the few old-timers who had voluntarily returned to China for a second tour of duty, were bitter. Our Old Man, who had helped the Chinese since 1937, had been given a raw deal by a small clique of generals. President Roosevelt, who had admired Chennault, was not any longer around to protect the former retired captain. When Winston Churchill met Chennault in President Roosevelt's Oval Office in 1943, he was quoted as saying, after the Old Man left the room, "How lucky we are that this man is on *our* side!" Now General Chennault, we were told, had "resigned" from China for some "medical reason" six weeks prior to the end of the war. Chennault was known to most Americans as "Old Leatherface of the Flying Tigers" and had become too much of a national hero. His reputation throughout the world made the two stars on his collar shine a lot brighter than the stars of some other generals in the China–Burma–India theater of war. What really burned us old China hands up, a few days later, was that our High Command didn't even have the decency to invite the General to the Japanese surrender ceremony on the U.S. battleship *Missouri* in Tokyo Harbor. Only General Douglas MacArthur asked: "Where is Chennault?"

Before his departure from China at the end of June 1945, I received a "Good-bye and good luck" letter from the General urging me to fly—once the war was over—only to Karachi in India, then to board a troopship in order to be "one soldier amongst thousands" when disembarking in the States. I would thus minimize the chance that some red-tape jerk, as he called him, would try to keep me out of America. The thoughtful General enclosed an unsolicited recommendation addressed to any possible employer for "after the war."

I was on the first list of men of the 14th Air Force to return to the USA together with my squadron commander, Major Lubner. On August 18, 1945, I said good-bye to China, my second home, which I did not expect to see again. A large tent city for thousands of GI's was being erected in the sand desert near Karachi's port. On a twenty-four-hour basis, loudspeakers blared out ranks, names and serial numbers of those who were to go home aboard the first troopship. It was a Victory ship already in port, in the process of being converted from freighter into troop transport to carry five thousand soldiers plus two hundred nurses instead of war supplies. We spent ten nights in the Indian desert near the docks, waiting for the conversion to be completed. Planning ahead, I bought in downtown Karachi a "typical" inlaid Bengal brass cocktail set with which —together with three coolie hats I had brought with me from China —I hoped to make a few American girls happy. Adding to the heavy brass my 120-bass Hohner accordion (on which I had played to AVGers and GI's, poorly but loudly, songs by Kate Smith as "When the Moon Comes over the Mountain" or the Ink Spots' "When the Lights Go On Again All over the World") which I had dragged with me around China since the 1941 AVG days, plus my German Plaubel camera and photo album, I had plenty to carry! Literally, that's all I owned; not one single piece of civilian clothes. . . . I had visions of a three-week sunshine cruise through the Arabian Sea, Suez Canal, Mediterranean, past Sicily and Gibraltar, across the Atlantic and back to the States.

Loading of the Victory ship began with a dozen men followed by the calling of "Master Sergeant Neumann, ASN 10500000." That's about all that went according to my cruise plan. Instead of us early birds having the choice in which cabin we'd like to be quartered, we were sent to the bottom of the ship; she was being loaded from the bottom up! Only as much baggage was allowed to be taken aboard by each soldier as he could carry from the ship in a single trip. To make matters even worse, a supply sergeant in Karachi had discovered several thousand Army winter coats that had arrived in steaming India by mistake. Ergo, each man was ordered to carry back home one winter coat in his duffel bag. The Navy lieutenant at the foot of the embarkation ramp sympathetically permitted me two trips to get my stuff down to my bunk on the lowest level of the ship. Cabins on deck were reserved for officers and nurses. My hammock, the lowest of three hung one on top of the other, was a foot from what turned out to be the ship's propeller shaft housing. Vertical space between the hammocks was so tight that my knees poked into the rear of the fellow above me whenever I turned over. For an unexplained reason, a few of the troops going home were assigned to serve as military police during the expected twenty-three-day sea journey. The bad news was that I was one of those picked for MP duty; the good news, that the top deck was assigned to me during the whole trip home. It was never made clear to any of us MP's what we were supposed to be watching out for.

The fifth deck below in our Victory ship was like an oven, even before we left Karachi. Outside temperature during the day hovered around 105 degrees F.; at night it went down to 90 degrees F. There was not a single porthole to get fresh air because we were below sea level. Once under way, we were fed two meals per day, standing at narrow counters. There was no place to sit down. The diet was simple: dehydrated milk, dehydrated eggs, dehydrated potatoes, dehydrated everything! (The ship's regular Navy crew, however, had fine quarters and three square meals a day.) We ran out of fresh wash water within four days after casting off from India. From then on there were saltwater showers only, which when mixed with soap left a sticky mess on one's skin. Seasickness began to afflict everyone shortly after departure; the foul smell in our hot quarters—four hundred men stuffed into the airless fifth deck below, with only two one-hour breathing spells on deck per day, to get fresh air—became unbearable. I swore to myself that I would never, never again take another boat ride.

The ship sailed under wartime blackout restrictions. Officers and nurses, of course, took full advantage of the blackout and the warm evenings in the many nooks, niches and under lifeboat covers aboard our transport. I didn't mind the social goings-on during the nightly blackouts, but what I did mind were the structural iron braces which held the ship's railing in vertical position. During my MP rounds at night (no flashlight was allowed) I had to feel my way along the rail of our rolling ship for four hours of duty at a time. My right shin hit those slanted, unevenly spaced braces exactly five inches below my knee. No matter how hard I tried to avoid them in the dark, I plowed into the sharp-cornered rods with painful regularity; it took the lump on my right shin three years to shrink back almost to its original size.

A one-day stop was scheduled at the city of Suez after passing through the canal, to pick up fuel, water and supplies. Just before our arrival, the captain's sarcastic voice, preceded by the traditional boatswain's whistle, came over the PA system: "Now hear this! This is your captain speaking. I understand that many of you are complaining about the service aboard. I apologize for the poor conditions, the lousy food, the unpleasant atmosphere in general. Here at Suez is your golden opportunity to get off this boat and make your own way back to the States. To those who can't stand it, I say, 'Please leave the boat.' To others I say, 'SHUT UP!!!' " Nobody left.

We had awful weather and rough seas the whole way home. Five thousand two hundred troops weighed little compared to a full load of cargo; the ship heeled and wallowed in breaking seas. High swells across the North Atlantic, once we passed Gibraltar, caused the propeller of our Victory ship to surface often, with consequent overspeeding of the shaft and rattling and shaking of the ship's hull, particularly the propeller housing. This was least appreciated by those poor devils—like myself—trying to sleep next to it. Several times en route home, the ship's crew held antiaircraft practice, shooting at balloons someone had launched three minutes earlier. How often did I berate myself for not having taken a chance on arriving in New York by air (Air Force personnel were entitled to *fly* home) instead of suffering through that miserable boat ride! But I praised the Lord for having put me in the Air Force during the past four years of war rather than in the Navy.

On the twenty-third day after leaving India, the PA system of our ship announced that we would arrive at New York late that night and would anchor outside the habor, awaiting daylight and high tide before entering port. The captain had been informed by the harbor authorities that many boats in New York were expected to welcome the troopship. Every one of the 5,200 troops would be permitted on deck, at the same time, to watch the spectacle as we sailed into the harbor. Then came what I feared most: a final reminder by the skipper that only one trip from each soldier's bunk to the dock would be permitted. "That's all. Good luck and God bless you!" We applauded in the spirit of reconciliation. After all, the captain *did* get us back safely to New York.

Finally, the moment we had been waiting for came: the last morning aboard the troopship. Everyone was up before the crack of dawn, heading for the top deck to find a space along the rail from where to view our entrance into New York Harbor. Two red-and-gold banners—FIRST TROOPS HOME FROM CHINA–BURMA–INDIA—had been fastened to the port and starboard bow railings. Knowing every nook and corner of the upper deck from my MP duty, I led twenty men to hidden stairs near the funnel, climbing up to a narrow platform thirty feet above top deck, just below the ship's foghorn. We didn't mind being squeezed at this finest of all viewpoints, high above everyone else!

Our ship raised its anchor at 0730, half a mile outside the harbor's entrance. Hundreds of private yachts motored out to greet us, big ones and little ones. Staten Island ferries and all freighters in the harbor displayed colorful signal flags. Fireboats played huge arcs of water. A WAC band on an excursion boat welcomed us with the German war song "Lili Marlene." Each ship as it passed, regardless of size, huge or tiny, let go with three "welcome home" blasts of its horns, whistles or bells. Our own foghorn, directly above us, answered each individual salute with three incredibly loud, deep-throated blasts! The air vibrated and shook so violently that we actually feared for our eardrums! There was now no way to get off that platform; every square inch of space below us, including the hidden stairs, was jammed solid. It was a most unforgettable morning. The sun broke through the clouds just at the right moment, shining on the Statue of Liberty, the Empire State Building and Manhattan's skyline.

Americans welcomed their boys home so sincerely that there were tears in the eyes of many of the toughest soldiers. Four tugs pushed our Victory boat up the Hudson River, then sideways against the Army pier in New Jersey.

Orders came over the PA system for all soldiers to return belowdeck and to stand by their hammocks while the first contingent of troops was disembarking, following the LIFO (last in—first out) system. Estimated time to get everyone ashore was two and a half hours. I knew that it would take a miracle to get me and my belongings off the boat in *one* trip. There were no stairs from one deck to the next. Instead, ladders consisting of iron rungs with vertical handrails on both sides led from the fifth to the top deck. Only one exit procedure seemed possible: I would use my pants belt to loop through the handle of the heavy accordion case and around the left side of my neck; three coolie hats would hang on my back with their thin chinstraps choking my throat. The bulky German camera would pull on my neck's right side. The duffel bag stuffed with the surplus Army overcoat, heavy Bengal brass, my photo album and whatever GI clothing I had left would have to be carried on my left shoulder, steadied by the left hand. I would use the right hand to pull myself up the five vertical ladders, grip by grip and rung by rung. This scheme required the help of one of my buddies, who would have to shove my rear end upward with his helmet on his head, so that I could grab the next few inches of the vertical rail as I struggled up from deck to deck. I reconfirmed my disembarkation strategy with the nice fellow who offered to give me a helping shove up the five ladders. After three hours of waiting came our turn: With a series of mighty pushes by the soldier below me, we made it to the top! I was a dripping mess of perspiration; my khaki shirt stuck to me like a wet rag; my neck was aching in each direction. . . .

On the dock below us a Navy band was playing continuously. I watched from above as Red Cross girls were handing each soldier something to drink and eat as he stepped ashore. I staggered along the deck toward the gangway, which was so narrow it could not accommodate the combined width of myself and the accordion. Quick action was imperative! Without asking me, the buddy behind me (who had shoved me up the five ladders) lifted the accordion over the wooden rail of the gangway, then let it drop, practically pulling my neck out of its socket. The heavy duffel bag on one shoulder, my free hand holding on to one rail, I tottered down the steep gangway, soaking wet and knees trembling.

On terra firma, a Red Cross girl stuck the straw of a milk carton into my mouth and walked along my left side holding the carton. I gulped down the milk while trying to catch up with the fellow ahead of me, walking as fast as he could to the waiting troop train for the discharge center at Fort Dix, New Jersey. Another ARC girl came with two sandwiches wrapped in wax paper, which were, it turned out, filled thick with strawberry jam. Since I needed both hands to balance the duffel bag, which had begun to slide off my shoulder, the sandwich girl pushed her gift inside my drenched shirt. Keeping my fast pace, I turned quickly to thank the two laughing Red Cross girls. At that instant the GI ahead of me stopped short because his duffel bag, too, had slid way backward and was about to hit the ground. I crashed into his bag. The inevitable happened: The collision opened the wax paper wrapping of the sandwiches in my shirt; jam dripped into my underwear, my pant legs, my socks and GI shoes. I dripped strawberry jam all the way to the railroad platform 250 feet away! At Fort Dix, an hour later, I was given top priority for a shower and to draw a dry uniform.

Every soldier arriving from overseas got a steak dinner on his first night on land, served by German prisoners of war; a quick medical examination followed in the morning. Each soldier could make one free telephone call to anywhere in the States. His service records were completed; he got his Honorable Discharge paper, and then came the final payoff: five cents per mile from Fort Dix "to the place of your enlistment." When my turn came for the latter, the disbursement sergeant wondered if he had heard correctly: ten thousand miles from Fort Dix to Kweilin, China? "Impossible!"

He called a major who made a Solomon-like decision: "We either fly you back to China, or you pick any place in the United States where you plan to settle and we'll pay you five cents per mile to there. Okay?"

"Yes, sir!"

No questions were ever asked about my citizenship. Chennault had been right. But my free phone call to Washington was a great disappointment: Clarice's roommate said that my lawyer friend Clarice happened to be in Buenos Aires at that moment. "No, she left no message behind for you. . . . No, I don't know when she'll be back." Oh well, I thought, there are other girls around. I'll head for California and see Mrs. Molthop's younger sister Bets, who had been writing me to China from Elsinore, California.

Discharged but still in uniform, I hitchhiked from LaGuardia Airport to San Francisco aboard an Air Transport Command plane. Fare: fifty cents, the rental fee for a parachute. I wanted to go to Los Angeles where most of America's aircraft plants were located. The plane, however, went to the Golden Gate City. In a Greyhound bus jammed with soldiers who had been discharged in the San Francisco area, twenty GI's and I stood for eight hours with our heads lowered until we reached Los Angeles. The bus drove along California's beautiful coastal highway 1A, past Carmel, San Luis Obispo and Santa Barbara. The country was beautiful! Our morale was great; we were singing and hollering like a bunch of kids. Beautiful Bets, acquaintance by correspondence, a model by profession, and sister-in-law of steel-plant owner Duane Molthop whom I had met in Chicago in 1944, met me at the Biltmore Hotel in Los Angeles. (Her husband, a B-24 bombardier, had been missing in action since 1942.) During the two-hour drive to Elsinore we became acquainted—and after one day of relaxing with her family in the eucalyptus grove surrounding their house, we felt like old friends. In the days to follow, Bets drove me daily to aircraft manufacturing plants near Los Angeles where I hoped to find employment; and also to clothing stores which possibly would have leftover civilian clothes that fit me—not an easy matter, since manufacture of civvies had been generally discontinued in America since 1942.

General Chennault's letter of commendation was good enough for the interviewers at three out of four aircraft plants to offer me a job, although just at that time thousands of workers and engineers were being laid off. I accepted a position as research and development engineer in Douglas Aircraft's Santa Monica plant. Bullock's in Westwood Village, a suburb of Los Angeles, still had the size of prewar clothes I needed—everything from top to bottom and from inside out. An efficient Douglas employee-relations girl, Vange R., located a bachelor pad for me next to the beach in the Santa Monica Canyon.

I was ready to report for work—my first job in America! But a hitch developed at the Douglas security office, which could not issue me a badge—without which I could not enter the laboratories. I was *still* an enemy alien! Douglas was aware that the House in Washington had already passed my naturalization bill, H.R. 3441, but that the Senate subcommittee was just then beginning to consider the matter. "Take a rest," Douglas suggested, and reassured me that no one else would get the job they had in mind for me.

Sitting around—waiting for an indefinite time—was not my style. There were hundreds of want ads in the *Los Angeles Times* for mechanics returning from the service. The majority of cars in America had been inadequately maintained during the war years and needed a lot of repairs. I was hired within minutes of my first interview by the Chevrolet/Oldsmobile dealer Bliss & Paden in Westwood Village. I bought a used bicycle to commute back and forth from Santa Monica, a pair of coveralls and a set of Craftsman hand tools at a nearby Sears. My first work in America began the following morning. The garage was open each day from 8 A.M. until 5 P.M., Monday through Friday. Although I was tempted to spend the weekends at the beach half a block from my rented room, I took an additional job over weekends at the CAC (California Automobile Club) garage from 6 P.M. Friday until 6 A.M. Monday. Not that I needed the extra income that badly, although I appreciated it and the always generous tips of those motorists who needed CAC's help, but because I was permitted to use CAC's tools and equipment—including their hydraulic car hoist. Over weekends I could maintain my own car—if only I had one! This didn't take long: For two hundred dollars I acquired a former beauty, a 1933 Pontiac club coupe. Its price was so low because only three of its eight cylinders were producing power and its body lacked every bit of paint. Four weeks later, the engine purred like a kitten and my paint job, with air for the spray gun coming out of the exhaust end of my landlady's vacuum cleaner, converted the rusty heap into a glossy car. I sold the Pontiac for six hundred dollars and bought myself the best auto I have ever owned, a baby-blue 1939 Chevrolet coupe.

Having my own wheels, I could now date to my heart's content! This, too, didn't take long. In appreciation for finding me the just-right apartment at the Pacific Ocean beach, I invited the PR gal of Douglas, Vange, for a couple of drinks (to which I was still not accustomed) and a dinner show at Hollywood's famous restaurant Carol's. (In the presence of my date I was asked in a booming voice by the somewhat insensitive cashier if I wanted "the good seats for fifteen dollars or the cheap ones for seven fifty"!) It was Vange's suggestion to see the lights of the glittering movie capital from above, after the show. Most bachelors knew the high Mulholland Drive off Sunset Boulevard, dead-ending at a cliff, and the parking strip along its edge from which one overlooked former antiaircraft searchlights crisscrossing the black sky. . . .

Other couples had had the same idea: We found just one empty space between cars facing the multitude of beautiful lights. Carefully, I steered my Chevrolet into the dark gap between two cars, the steep abyss before us, and stepped on the brake—but my Chevy kept moving, very slowly, toward the vertical drop only a few feet away! I pushed with my foot as hard as I could; in addition I used all my strength to pull the handbrake—but still the car kept rolling toward an inevitable disaster. I broke out in a cold sweat. "Jump! Jump!" I shouted in desperation to Vange—and she did! Only then I noticed—to my enormous embarrassment—that the car to my left had been backing out at precisely the moment I had come to a stop. The optical illusion, making one think that one's own car is moving, is common. Cautiously, my date got back into my car; I felt like an absolute idiot as I drove her home in total silence.

Still angry with myself, I reached my room much earlier than I had wanted to. My British landlady had placed one of her baking miracles, a small chocolate pie generously sprinkled with jimmies, at the foot of my door in the hallway. Normally, such a creation would have exerted an irresistible temptation right then and there; but not this evening. My appetite was shot. I placed the pie on my table, switched off the light and went to sleep. At two in the morning I awoke. Remembering the pie, I stepped out of bed with one leg, reached over to the table, pulled the plate with the pie to me. I sat up in bed and began to eat in the dark. Halfway through the pie, I felt some odd movement on my cheek, then on my neck, seconds later on my tongue and deep down in my throat . . . I stopped eating, trying to analyze what it could be. I switched on the light—and saw a swarm of black ants rushing in all directions over my bedsheets, myself, my hand, my pie. I jumped out of bed, ran into the bathroom and began to spit . . . dead and live ants coming out of my mouth. I must have eaten hundreds of them. This, I can attest, was not one of my better evenings.

At the Chevrolet garage in Westwood, I learned much about American civilians: the simpler they were and the older their cars, the nicer and more appreciative they were. It didn't matter whether they were schoolteachers or farmers: Without fail, the simple person would make a special effort to locate the mechanic who had fixed his or her car and to give him a quarter, with a friendly "Buy yourself a beer!" The wealthy owners of big La Salles, Buicks or Chryslers disappeared from the scene without as much as a thank-you.

Three nights a week I attended the engineering refresher course at U.C.L.A. (University of California at Los Angeles) under the GI Bill of Rights. Not only did I have to relearn what I had been taught in Germany seven years earlier, but all equations and formulae which I had learned in Germany in simple metric had to be modified into the complicated and difficult-to-apply inch system. Just in case an unexpected snag should develop before I started to work for Douglas (e.g., in case the Senate or President would not pass the bill granting me U.S. citizenship), I took the Federal Aviation Agency's A & E (aircraft and engine) mechanic's course and test, and then got my FAA license. I could always try to get a job as an aircraft mechanic working for an airline.

But nothing went wrong. The Douglas people kept track of my bill's progress in Washington. After the Senate had passed it, with President Truman's signature only a matter of time, they asked me to report to work at once. I kept the weekend job at the California Automobile Club to add to my savings. One Saturday morning at the garage, I received a call from the owner of a Packard limousine who said, "Battery is dead." He was waiting for me in the empty parking lot of a Santa Monica bank. I found a tall, lanky, gray-haired man leaning against the black Packard's front fender, his arms folded. A little lady sat in the rear. Looking for his dead battery, I lifted the right side of the Packard's engine hood (that car model still had a piano hinge atop). All I saw was a vertical wooden board blanketing the side of the engine, with a large number of horizontal pipes of increasing lengths, one above the other. I quickly lowered that side of the hood, walked around the Packard and lifted the left side: another vertical board blocking the view from bottom to top of the engine, but this one was cluttered with wirings and solenoids. Next, I knelt on the ground to peek below the floorboard on the driver's side: no battery! What was all this? I looked up at the man who was watching me.

"I am Boris Karloff," he introduced himself.

"I'm Gerhard Neumann," I replied.

"Don't you know who I am?"

"No."

"Didn't you ever see *Frankenstein*?"

"You mean the castle? No, I've never been at the Rhine."

The tall man looked incredulous and his size seemed to shrink before my eyes. He then explained the mystery of the missing batteries. An electric organ powered by four batteries located under the rear seat, two on each side of the car's differential, was installed in his Packard. The organ's keyboard was below the glass partition between driver's seat and rear compartment. The little lady, Mrs. Karloff, had been playing the organ while her husband spent a long time in the bank—and had run down the batteries. After I started the Packard, Boris Karloff asked me if I could read a German car maintenance manual, and invited me to visit his home in Beverly Hills, where he had "the world's best car," a Mercedes-Benz 300 SSK, which he would not bring into *any* garage, under any circumstances. All his car needed was a careful checkup, he said. He had many friends, actors and actresses, who also owned fancy foreign cars and who would be delighted if I would check their cars in their homes on Saturdays, he added. Shades of China! I discontinued work at the Automobile Club garage and began to make the rounds to some of Hollywood's famous movie stars' homes, including Ingrid Bergman's. My self-established limit was two house calls every Saturday for two months. From then on, I spent Saturdays and Sundays at the beach.

And what happened at work at Douglas? It was most satisfying. I learned to work with technical associates, to carry out experiments in an advanced-technology atmosphere, to write and dictate technical reports. I worked sometimes days, sometimes nights. Assignments included research on fog particles and their effect on airplane controls; cabin supercharger shaft failures; jet engine afterburner development (at that time the responsibilities for that jet engine component were with the airplane rather than the engine builder). Mr. Donald Douglas, senior, whom I met twice, the company's management, my bosses, peers and even my cute blond secretary (whom I shared with half a dozen other engineers) were all splendid.

Then out of the blue came a surprise: a cable from General Chennault, from Shanghai. He offered me a position as his special projects engineer in an airline he was in the process of forming in China. He already had bought two dozen war-surplus, brand-new C-46 transports which were sitting in Hawaii. His salary offer was a multiple of the pay I got at Douglas, tax-free. I acknowledged my interest and cabled to the General that I'd let him know soon.

After some cables back and forth to Germany, where she was then working as an attorney for the U.S. Department of Justice, Clarice flew across the Atlantic to New York and Washington in October 1946, then on to the Los Angeles airport at Burbank. It was Clarice's first visit to the West Coast. The weather that day was beautiful. We had two bananas in the car to eat en route to Santa Monica. I was so pleased and bewildered to see Clarice again, I threw away the banana—and kept the peel in my hand. . . . We were married forty-eight hours later.

I still could not believe it: the lawyer from Washington and my first dinner date in the USA—the pretty little bundle of energy! Clarice, after a quick flight back to her home in Connecticut, drove her DeSoto coupe back to Santa Monica; in it traveled a six-month-old, Berlin-born Airdale terrier, Mr. Chips. I gave notice to Douglas (which graciously offered to rehire me whenever I returned from China) and cabled General Chennault that there would come two Neumanns plus a dog. I sold my beloved Chevy and began, together with Clarice, to make lists of what we should pack into the DeSoto and what into a few crates to be shipped to China. The list included dog food, sleeping bags, army cots, dried coffee, two wonderful Halliburton aluminum suitcases packed with winter and summer clothes, and a Klepper *Faltboot* which Clarice had sent me from Germany. The DeSoto, we were assured by the shipping agent, would be loaded belowdeck—protected from the elements —on a slow boat to China and would leave San Francisco in one week.

In January 1947, I left on a United Airlines DC-4 for Honolulu. We waved good-bye to each other—"See you in Shanghai!" The following day Clarice and Chipsy boarded the former troopship S.S. *Marine Lynx* in San Francisco.

12

Back to China . . . with My First Date in America

In 1947 there were only three hotels along Waikiki Beach in the Hawaiian Islands. The crew with which I was to ferry to China one of the twenty-four new C-46 freight planes General Chennault had bought on the war-surplus market for his newly formed airline CNRRA (Chinese National Relief and Rehabilitation Airline), later to become CAT (Civil Air Transport), stayed in Hotel Moana. One thirty-minute checkout flight around the island, past Diamond Head and over Pearl Harbor, was all we made before declaring the plane ready for the long trans-Pacific flight to CNRRA's main base at Canton, with several stops en route. It had been less than six years ago that the Japanese surprise attack (which should not have been a total surprise to the USA) forced the United States to declare war on Japan.

Twin-engine planes like our C-46 had always been prohibited by international civil aeronautical authorities from flying over wide expanses of open water—in case one of the engines failed. The reason was simple: Depending on the weather and even if all air cargo was jettisoned, a single propeller engine might not be able to keep the plane from losing altitude and its ability to reach an airfield on the coast. General Chennault could not care less for the regulations. To the contrary: He had Hawaiian mechanics install two temporary fuel tanks in the middle of each C-46 fuselage, to increase our overwater range. He did not want us to run out of gas on the way to the Orient, in case we'd get lost somewhere over the wide Pacific! The flight plan for each plane, one leaving every day, called for the first overnight stop at Johnston Island. This was a rock the size of an aircraft carrier deck, barely sticking out of the Pacific Ocean. Nothing was on it but one Quonset hut and a population of a thousand gooney birds squatting in the center of the warm black asphalt strip. An arriving plane had to scare the birds off the runway by buzzing them several times before being able to land. To clear the strip for takeoff next morning, rifle shots had to be fired to shoo the birds into the air.

We made overnight stops at the islands of Kwajalein, Guam and the Philippines. Rusty hulls of shot-up World War II landing craft and American and Japanese transports were still lying half submerged near the invasion shores or farther out where they had been sunk. Some had run aground on coral reefs clearly visible from the air in the sparkling light-blue water. After three days of rest and sightseeing in Manila over Easter, we flew on to Canton in South China, eighty miles north of Hong Kong. For Chennault's married employees, his staff had rented two-story houses in Tungshan, a suburb near the airfield. Until Clarice arrived, I lived alone in an upstairs flat. Three weeks later I flew north to welcome her and Mr. Chips to China, on the waterfront of Shanghai. It was a thrill to see Clarice waving to me from the deck of the S.S. *Marine Lynx* and to watch her and Mr. Chips walk down the steep gangway.

After a few days in Shanghai's Park Hotel, and after Clarice and Chipsy had had a chance to mingle with masses of Chinese and take their first rickshaw rides, we three flew back to Canton. The American couple occupying the floor below us had just decided to get rid of some large rats residing at the bottom of our joint chimney. The rats had shown a special interest in the couple's food storage. Consequently, the husband, one of CNRRA's pilots, blocked their kitchen flue with bricks. With the rodents' entrance prevented into the kitchen downstairs, they climbed inside the chimney to the second floor and got out at the Neumanns' kitchen, where they promptly took a liking to whatever Clarice and the amah bought daily at the open-air market. The rats, one-third the size of dachshunds, scurried brazenly around our kitchen in broad daylight. One day, a tug-of-war developed when the biggest of the rodents pulled on one end of a large fish just brought home from the market while my wife hung bravely onto the fish tail. That was enough: We placed water-filled pans under each kitchen table leg to keep the rats off its top. We hung our bananas from the ceiling-lamp cord. We, too, barricaded our flue, which had been used by our amah to work the clay stove properly. From then on, an open window in the kitchen had to provide the draft for the amah's charcoal fire. The stove sat on the floor, with her squatting on her haunches in front of it and waving a bamboo fan back and forth.

Cockroaches were other uninvited guests in our Canton flat; we battled them daily in disgust but finally gave up. In the evenings, we would reach around doorframes to switch on the light before entering a room, kitchen or bath, and gave the roaches half a minute to scurry under the furniture or tub. If they didn't take advantage of this time grace, they were liable to be squashed on the tile floors with a loud crack underneath our shoe soles.

The weather in Canton was hot and so humid that mold formed on leather shoes and belts within one day. Our amah, who understood no English but had been bright enough to have had an American boyfriend during the war in some other part of China, was a joy. When she learned that I, too, had served during the war in China, she asked me if I knew her beau.

"What was his name?"

Her face beamed from ear to ear. "G.I. Joe!"

After one month of riding to and from the dusty airfield on an open-bed truck, we heard that Clarice's DeSoto had arrived at a Shanghai dock. She and Mr. Chips flew there on a CNRRA company plane and began the battle to get her car through Chinese customs—without paying the bribes normally expected by the officials. I arrived at Shanghai a day too late to assist Clarice, but in time to share her dismay when she spotted the rusted DeSoto that had been an attractive, glossy maroon coupe when we saw it last on the loading dock in San Francisco. Obviously, the car had been exposed to saltwater spray on the freighter and had not been stowed belowdeck, as had been promised. It also had been broken into: Four cases of dog food had been stolen. There was nothing we could do but locate a car painter. Duco spray paint was available in Shanghai but no color sample charts. The Neumanns and the paint-shop owner stood for twenty minutes on the curb of Shanghai's busiest street, Bubbling Well Road, to watch the color of cars zipping by. We pointed out to the painter which color we wanted and contracted for a "first-class pearl-gray lacquer job." Six ten-year-old helpers would do the hand-polishing.

At that time the airline asked some of us to transfer to Shanghai, where General Chennault was planning to move CNRRA's main base of operation. Before the war, Shanghai was the world's largest city and much more cosmopolitan than provincial Canton, in which no English was spoken in spite of its proximity to British Hong Kong. In Shanghai proper lived over eight million inhabitants; in Greater Shanghai twelve million; thousands were White Russians, thousands were Jewish refugees from Hitler's regime, thousands were foreign businessmen busily selling their wares in a city starved for anything Western since the beginning of the Sino-Japanese War in 1937. The Bund was, and is, Shanghai's most beautiful street; it is a tree-lined avenue alongside the colorful harbor crowded with freighters and sail junks. Located along the waterfront were foreign embassies and consulates. Near the Bund were modern stores with escalators, restaurants, roof gardens and everything the biggest city of the world ought to have. And just as in boisterous New York, where one can find small, quiet churches next to skyscrapers, lively Shanghai had next to its high-rise buildings little temples. There Buddhists prayed quietly and lit incense sticks. Shaven-headed monks wearing orange robes and straw sandals rang tiny bells, and every so often beat wooden drums hanging from the ceiling.

After two years of peace in China (1945–47) one could buy anything in Shanghai, from the latest American car to German electrical appliances. It was also the sin city of the world, crowded with nightclubs and dance halls; it had modern shopping centers, a racetrack in the middle of town, gambling casinos, millions of honest workers, lots of Chinese and white crooks, thousands of taxis and rickshaws—but very few telephones in 1947.

I supervised Chinese mechanics who serviced, on an around-the-clock basis, the twenty-four C-46 transports that had arrived from Hawaii and were now busy supplying the Nationalists. Because the American parts manager was drunk almost every day, Clarice helped out to keep the supply line moving and CNRRA's planes flying from Shanghai. She organized a warehouse for aircraft and engine parts and set up a spare-parts card-filing system. At first we had very few parts, then received many more as surplus crates with parts for C-46's and their Curtiss-Wright engines began to arrive from the former American bases on Okinawa and Saipan. I can still see Clarice, standing in her oversized brown coveralls next to opened crates, comparing newly arrived parts with the technical sketches in C-46 maintenance manuals. Chinese carpenters were hired to build storage bins to her specifications.

Each evening, seven days a week, when we returned home from the dusty, hot Hungjao airfield, one honk of the horn in front of our garden gate caused it to open miraculously. Our houseboy, about fifty, spoke English fluently and was able to read our minds. He waited there in his spotless, starched white uniform, on the driver's side of our company's Jeep holding a chromium tray with two large glasses of ice water! Clarice and I often wondered why such an intelligent, bilingual Chinese would work for the relatively low pay he got from us. The mystery was solved when Clarice stayed at home one day: An hour after we usually left for the airfield, until an hour before we came home, our phone rang every few minutes; Boy made brief calls himself after he had answered a ring. Clarice discovered that Boy owned a fleet of rickshaws and pedicabs, and he made a healthy profit dispatching them from our house thanks to the day-long availability of a phone at the Neumann household.

Many ways to make a quick buck were practiced in the China of 1947: The official exchange rate of a U.S. dollar to the Chinese yuan, for example, was different in Shanghai than it was in Hong Kong, usually by 1 to 2 percent. Someone could—and did—just shuttle back and forth between these two cities exchanging, perfectly legally, his currency in one city and selling it in the other, making thousands of U.S. dollars in the process.

Chinese officials in Shanghai had more get-rich-quick ideas. One evening I received a phone call asking whether I'd be interested in designing an automobile directional traffic indicator that could be manufactured locally. Such a gadget had to be attachable to the outside of any car. The anonymous caller stated that the scheme was for the chief of police to invest personally in the production of a large number of these indicators; when an adequate quantity was available to equip each car and truck in Shanghai, the chief of police would issue a municipal ordinance directing that, by a certain date, all motor vehicles in Shanghai would be required to have traffic indicators. The time span between edict and installation was to be made so short that no one could compete with the indicator manufactured on behalf of the police chief.

Another Shanghai official planned to get rich quickly by having thousands of cotton shirts made, then distributed to police stations around town. An ordinance was to be issued requiring each rickshaw coolie to wear a shirt "while working on public roadways." The official reason for the edict: "Rickshaw coolies shall henceforth not offend the public by going topless. . . ." If the coolie did not own a shirt (most of the coolies owned nothing but a bamboo hat, a towel around their neck, a pair of shorts plus straw sandals) he would be taken to the nearest police station, where there happened to be available—for a price—the mandatory cotton shirt.

U.S. Secretary of State George C. Marshall's efforts to mediate between Mao Tse-tung and Chiang Kai-shek in 1946 had been unsuccessful. Now, in 1947, one could tell from CNRRA's flight plans that the Communists were making progress in their southern advance from their hideouts in northern China. Instead of civilian goods, we began to fly more and more weapons and ammunition to support the Nationalist troops fighting north of Peking, and returned with wounded soldiers (or very wealthy refugees who could afford plane fares). The condition of our C-46's deteriorated in direct proportion to the intensification of the Civil War. We began to have less and less time to perform scheduled maintenance on the airplanes. Once again—just as in World War II—we had to make do. We cannibalized a C-46 hangar queen for parts and moved instruments quickly from one plane that had just returned and installed them into another one ready for takeoff. Fortunately, we lost no planes to enemy ground fire, but we had one particularly memorable incident: One of the two engines of a C-46 taking off from Hong Kong's Kai Tak Airport "froze" (for lack of clean lubricating oil) shortly after it had climbed to 5,000 feet on the way to its cruising altitude of 20,000 feet. The propeller stopped dead. To maintain altitude on only one running engine and be able to return to Kai Tak, the crew had quickly to lighten the plane by shoving heavy boxes out of the cargo door. The crates dropped into rice paddies and scattered. Over half the load had to be dumped before the plane was able to maintain altitude with a single engine running. The problem was not only that we had to drop cargo overboard, but that these crates were full of brand-new bank notes. Several billion yuans! A military unit was immediately dispatched to find the crates; by nightfall it claimed to have recovered less than half the money—spread over a wide area.

When it became obvious that the Communists' march from the north of China was unstoppable, that they had popular support and their arrival at Shanghai was only a matter of time, Chennault decided to transfer our main operating base back to Canton in the south. Once again we had to move. Clarice's DeSoto was put up for sale in Shanghai. The demand for reliable American automobiles grew as the Communists continued their advance. Selling her 1940 car was not difficult, but *how* to get paid for it was! Everyone claimed not to own any U.S. dollars, whose possession had been outlawed in China recently. She received offers of platinum, gold bars, diamonds, jade and Chinese cash (which daily lost more and more of its value). We *insisted* on receiving U.S. dollars for the DeSoto; ultimately, a Coca-Cola company executive wanted Clarice's car badly enough to deposit U.S. dollars from his Swiss bank into Clarice's account in the States.

The evening before we left for Canton we visited with General Chennault, who had decided to remain in the big city to the very last. In his home we met charming Miss Anna Chan, born in Peking and educated in Hong Kong. (Just before Christmas 1947, she became Mrs. Claire Lee Chennault.) I had seen her once before, toward the end of the war, when she was an eighteen-year-old war correspondent interviewing the General. (Today, Anna Chennault is still beautiful and a prominent person in her own right in Washington, a friend of the U.S. President, a high-level representative of our government and a society hostess. As a consultant to major companies, she specializes in countries in the Orient, from Korea to the Philippines. In addition, Anna Chennault is the author of a dozen books including a best seller.)

We located a nice home for rent at Canton's outskirts and had an electrician come in to check out the 220-volt electrical system. He gripped the bare ends of the hot and ground wires, one in each hand: The 220-volt current flowed through the wires and his tough body. The pomaded hair on his head stood out like the quills on a porcupine. If his hair had stayed smoothly on his head, we (and he) had a problem! Who needs a voltmeter?

We were just beginning to settle down in that new home, the third in six months, when word spread that even Canton would have to be evacuated soon; the renamed airline CAT was going to purchase a barge, equip it like a workshop, then tow it to Taiwan, formerly Formosa, where the Nationalist government had moved. The Neumanns, however, decided to return to the States. I wrote to General Chennault in Shanghai, giving him notice. He answered promptly that he had felt for some time that my "engineering training and experience could be better utilized in the States." How to get out of China was up to us. It was not as simple as we first thought: Neither civilian ships nor airlines had resumed regular operations from the Orient to the States via the Pacific; airline tickets via Europe were too expensive.

As I discussed these difficulties with Clarice at lunch in Canton, she casually asked, "Why don't we drive home?" What? *Drive* home? How on earth could we drive home from China? "Via Europe or North Africa" was her answer, "then by boat back to the States." I knew that Clarice was not familiar with the territory between China and Europe across the Asian continent: Siam, Burma, India, Afghanistan, Iran and the Mideastern countries through which we would have to travel; she could not have considered the difficulty—if not impossibility—of obtaining gasoline when and where we would need it, now that the war and its many sources of supply were becoming rapidly a thing of the past. I knew that there were hundreds of miles of jungle and desert without any roads whatsoever, or places to stay, or water to drink . . . just wilderness, jungle or mountains and snow. *Drive* home? Impossible!

But then—Within the next hour I began to warm up to her idea. By Jeep? With four-wheel drive? With enough five-gallon jerry cans for gasoline? With snow chains, sleeping bags, a tarpaulin for a tent, and tools? With a seat for Chipsy? I got up from the lunch table, found in my bag an Esso map of the 1942 World War II area covering China and Burma to the Indian border. The only other map I had was a 1946 National Geographic political map of Asia, which pictured that large continent and Asia's little appendix,

Europe. On this very small-scale map was shown a thin red line weaving from Iran to China, with a footnote that Genghis Khan had traveled this trail 750 years ago. If Genghis without a four-wheel-drive Jeep did it, why couldn't we? I changed my vote and concurred with Clarice's "Why don't we drive home?"

With this decision made, the next steps became obvious: First, we would have to find a suitable Jeep; then obtain transit visas through the many countries we would traverse. For the most part, the former colonies of France and Britain were in a state of political turmoil. Britain had promised her colonies in 1942, as incentive for their support during the war with the Japanese, that they would be given their independence two years after war's end. Now that the British were beginning to leave, native factions began promptly to compete with each other for political control of their countries. If we really wanted to *drive* out of the Orient, we would have to do it very soon because of the approaching winter weather in Afghanistan. I had read somewhere that communication among cities in that mountain state was at a virtual standstill between December and March. We would not get going until the end of the rainy season in southern Asia (which causes muddy tracks and flooded streams), yet we had to arrive in Afghanistan before deep snow would make crossing that mountainous country impossible. We had to make lists of what to take with us in the very limited space of a Jeep. There would be few opportunities en route to purchase clothing, food, Jeep spare parts or tools.

We booked two seats on that afternoon's flight from Canton to Hong Kong on CNAC, the airline that had helped me slip out of His Majesty's Crown Colony when I was interned there in 1939–40. When we got to the airport for the twenty-minute flight, passengers in the converted military transport were strapping themselves to their bucket seats along both sides of the cabin. Two women amongst the thirty Chinese passengers aboard were already airsick, even before the plane's engines had been started. But our departure for Hong Kong that afternoon was questionable: The American pilot noticed that his C-47 had been leaking hydraulic fluid; he was debating whether or not to cancel the flight. The captain would do so only reluctantly, he told me, since he had a "Suzie Wong" date that night in Hong Kong. I then introduced myself to him and, even though I was still working for a rival airline, offered my help as an FAA-licensed airplane mechanic and told him that my wife and I were just as anxious to get to the British colony as he was. The CNAC captain accepted. In two hours of work rerouting pipes and bypassing the leak in the hydraulic pump, I had his C-47 ready to go with the proviso that the pilot would fly it with its landing gear in a down position to assure a safe landing in Hong Kong. The plane had already lost too much of its hydraulic fluid, and what was left would be needed for brakes.

Several used Jeeps were listed for sale by the Royal Air Force in the ad section of Hong Kong's daily, the *South China Morning Post.* "Tenders are invited," it stated. Half a dozen Jeeps, some with their bodies wrecked but with good engines, others with engines worn out but with undamaged bodies, were for sale. Each Royal Air Force Jeep had large red-white-blue circles painted on its hood for aerial identification. I drove each Jeep that was in running condition up and down the steep hills of Victoria. None was suitable "as is," but a combination of a good chassis with a fine engine would make a perfect vehicle! I bought two of the RAF Jeeps, both made by Willys: one with a good body, the other with a good engine and a good transmission, for U.S. $400. Also, five new six-ply Seiberling tires for a total of $100.

Claude White, the American who had employed me in his Far East Motors garage eight years earlier, in 1939, and who himself had been interned for three and a half years by the Japanese in Hong Kong's Aberdeen Prison, was in good spirits. White was unperturbed by our decision to drive to Europe and offered his garage for any work I wanted to have done or, as he suspected, wanted to do myself (the jungles of Siam or Burma were not exactly places where one wanted to experience a mechanical breakdown).

When we applied for transit visas through French Indo-China, France firmly refused to give us permission to drive in Indo-China because they "could not guarantee" our safety along Coastal Highway #1, between Hanoi and Saigon. The Thai, British-Burmese and British-Indian consulates, however, stamped visas into our American passports while we waited. Butterfield & Swire, the Orient's giant shipping company, told us that one of its British coastal freighters would be leaving for Bangkok in ten days and assured us—if we somehow could get the Jeep "alongside the S.S. *Newchwang*" moored in the middle of Hong Kong's harbor—the Jeep could be loaded without any problem. How about ourselves? The ship had no passenger accommodations. If we were willing to sleep on deck, that would be okay with Butterfield & Swire. No charge for us and the dog.

Mission accomplished, we returned by train to Canton and packed clothes, cots, sleeping bags, and tools, which we had brought from the States nine months ago. When we had bought the stuff in Los Angeles war-surplus stores, we, of course, had not the slightest idea that we would ever use it on a high-risk, extraordinary trip across Asia. Our amah was in tears; we left her the Ping-Pong table a carpenter had made for us (she was going to use it as a bed for herself and her child) and an extra month's salary. From Canton's telegraph office we sent a rush cable to Haile Selassie in Addis Ababa (Ethiopia) offering my services with the Emperor's new airline, which, according to the latest *Reader's Digest,* TWA was going to operate for him and which needed engineers. We asked Emperor Selassie to answer by return cable care of Butterfield & Swire in Hong Kong.

For the third time I said good-bye to China. Clarice, Mr. Chips and I boarded a riverboat traveling from Canton to Hong Kong, together with the baggage we were going to take across Asia. In his last letter General Chennault had warned us, once more, of the tense Mideast political situation, which had grown touchy because of the American proposal before the United Nations to partition Palestine in 1948 and to create a Jewish state. As a consequence, America's reputation was terrible in the Arab Mideast. Personally, I was more concerned about any Jeep spare parts I might need beyond the usual fan belt, four spark plugs, and ignition breaker points. We would have to buy a tire repair kit, a good tire pump, a truck jack, and a 20-foot-long rubber hose to suck gasoline out of jerry cans positioned high on the Jeep's rear toolboxes and to transfer the gas into the Jeep's tank. We decided to keep the better of the two worn-out batteries since a new one was too expensive; I didn't mind cranking the Jeep by hand each morning to get it started.

The range of our Jeep with its twelve-gallon tank plus the forty gallons I intended to carry in eight U.S. Army jerry cans (their filler necks pointing to the outside so that I could siphon the gasoline out of each can without having to unload the Jeep) was about 1,100 to 1,200 miles. Within such distance, I thought, there ought to be someplace where we would find gas to refill our capacity. A ninth jerry can would be filled with five gallons of engine oil and a tenth with drinking water. The total length of the journey—as we judged from the very small-scale map we had—would be about 10,000 miles, including major detours and the many S curves to crawl over the high mountain passes. The rainy season was just ending; roads and tracks should be drying out quickly, in an adequate condition for a four-wheel-drive vehicle. We would eat whatever the local population ate. And Airedale terrier Chipsy? He had to do likewise. "When in Rome, do as the Romans do," we told him.

Working five days in White's garage, I made one Jeep out of the two we had bought by transferring the good engine and four-wheel-drive transmission into the good chassis. I had Chinese mechanics mount the new tires, then test-drove the finished product extensively. October 7, 1947, was the day when a sail junk was hired to pick up the Jeep at the Yaumati dock in Kowloon and deliver all three of us to the freighter S.S. *Newchwang,* moored in the middle of Hong Kong's large harbor. A typhoon warning that morning—three red balls hoisted up a mast at the Kowloon police station, reminding me of China's air raid warning signals—delayed the ship's departure by one day. We left on my thirtieth birthday, October 8, 1947. There was much yelling, pushing, swearing in German and Chinese as, holding my breath, I slowly drove the empty Jeep over two narrow, flexing wooden planks—inch by inch—from the high quay onto the deck of the much lower junk across fifteen feet of open water. Clarice and Chipsy joined me on the boat. The all-female junk crew (five women in their twenties, dressed in black silk suits and their hair in long pigtails) hoisted bamboo sails; the old woman owner manned the rudder. Our historic trip got under way! We had christened our Jeep *Rosinante,* remembering the story—and the heroic horse—of Don Quixote.

Four British officers and forty Chinese crew members of the S.S. *Newchwang* watched as the strange cargo for Bangkok arrived under sail. Another few breathless moments came as *Rosinante* and Chipsy were hoisted in a giant net aboard the freighter. Clarice and I then had to climb up the Manila rope boarding net hanging vertically from the ship's deck, the Hong Kong harbor forty feet below us! The captain, officers and Chinese crew gave us a cheerful welcome. We had time before departure from Hong Kong to go once more ashore in the *Newchwang*'s motorized lifeboat, to visit the cable office on Victoria hoping to find a reply from Emperor Haile Selassie in Ethiopia. Yes, we *had* a return telegram from Addis Ababa, at the very last moment. I tore open the envelope. The cable stated in English that ours from Canton could not be delivered because there were several Haile Selassies in Addis Ababa. "WHICH HAILE SELASSIE DO YOU MEAN?" it asked.

The British officers of the *Newchwang* offered Clarice and me their own quarters for the week aboard, but we preferred to sleep the first night on our cots, under a tropical star-studded sky. It was a relaxing trip with the Britishers, who could not do enough for us two Americans whose country they gave full credit for their rescue from Japanese prison in Manchuria in 1945. Clarice and I played chess, read, wrote letters and promised poor Mr. Chips that there would be plenty of trees once we got ashore in Thailand.

Immigration formalities were quickly completed five miles outside of Bangkok where the *Newchwang* anchored. We were told that there were only three hotels in all of Bangkok in 1947: the very simple Hotel Europe, a mediocre one and the high-class Oriental, where I had stayed eight years earlier en route to China. The Europe was just right for us three: very low cost; on the second floor, a bare bedroom with the standard electric ceiling fan, mosquito net over each bed, a washbasin that had to be emptied out of the window onto a side street, and a sign in English: LADIES, PLEASE REMAIN IN THE BEDROOM. Clarice commented, "Keep your noses to the grindstone, ladies." It had neither bathtub nor shower, but a large earthen jar filled with water and a hand ladle to pour water into the adjacent toilet. A huge boa constrictor in a large cage below our window, and a chained large monkey showing off his antics in a tree opposite our open bedroom window, completed the decor. *Rosinante* remained at the docks the first night.

Our embassy provided us with an interpreter to get *Rosinante* through customs. This was a complicated matter since no customs officer believed our story that we were going to "transit" Thailand: How could we transit if there were no roads? they wanted to know. The local officials suspected us of importing the Jeep for sale in town—without paying import duty. Then we faced Problem #2: "Where are your license plates?" We had none; I had never even thought of needing any. A whole day was wasted at the Bangkok police station, which issued license plates for the few motor vehicles in town but never before had given any plates to a Jeep. This raised

Problem #3: The barefooted police officer in charge of licenses insisted that we have a directional traffic indicator, on the Jeep's side opposite the driver. This was total nonsense, of course: Only a handful of cars operated in Bangkok; besides, we were going to leave in a few days anyway. But the police officer persisted; he wanted us to have *some* directional indicator on the right side of the Jeep. I strained my engineering brain how easiest to rig up some blinking light on the right fender. Clarice, in the meantime, tied a whisk broom with rubber bands to the end of the long Jeep engine crank and put this assembly behind and below the windshield into two brackets originally installed to hold a rifle. Pushing the crank handle to the right, she caused the whisk broom to emerge beyond the windshield. I told Clarice that her idea was ridiculous, but the Bangkok police laughed and were delighted—and issued us two Siamese license plates for *Rosinante*'s front and rear bumpers, free of charge.

By sheer luck I noticed on the sandy floor, in a corner of the police headquarters, a pile of white, oval metal plates with the black letters *SM.* Such international plates, with appropriate letters on them, are issued by any country's automobile club and are attached to bumpers or body whenever a member car is temporarily driven abroad. I asked the police chief how we could obtain two such plates and the triptyque (threefold) international document that went with them. "Take as many plates and documents as you want," the chief told me through an interpreter, and he pointed at the pile on the windowsill. No charge! The reason for his generosity was simple: The Royal Siamese Automobile Club had been discontinued since Japan began to occupy Thailand in 1941; plates and the accompanying triptyques with their empty spaces, even if filled out, were therefore invalid. All printing in them was in Siamese. Only 20 percent of all Thai people could read Siamese; certainly no one outside Thailand!

Clarice and I went to work in our hotel room and filled out each empty space on the front page of the booklet, to make it *look* official. I carefully drew on one dotted line the logo known worldwide *Ford* (although our *Rosinante* was made by Willys); on two other blank spaces I filled in the chassis and engine numbers as they appeared on one of the two bills of sale from the Royal Air Force. The booklet had a total of twenty-four pages, including the one on which we described our Jeep; the other pages were for foreign countries to stamp our Jeep's "in" and "out" passages. (As it turned out, we passed through ten countries on this journey. Engine and chassis numbers were examined by each country's border guards; each time we received an "okay" with impressive-looking stamps and signatures entered into our worthless Siamese Automobile Club booklet. And, of course, the engine serial number was wrong, since it came from the engine we had left behind in Mr. White's garage in Hong Kong.)

A Bangkok tailor modified *Rosinante*'s canvas top so that we could drive the Jeep partially open, like a phaeton, or—when the weather was cold—like a closed box with its side curtains tied down. I asked a blacksmith to insert a flexible section into the exhaust pipe and reroute it to give the Jeep more ground clearance: Instead of emerging underneath *Rosinante*'s left rear, the exhaust pipe led across the Jeep above the floor, and emerged on the right side. Although a bit too warm for us in the jungle, it turned out to be truly a lifesaving heater at 10,000-foot altitude, with howling winds and a foot of snow. The blacksmith also attached a U-shaped handgrip to the right side of the dashboard, to support the bouncing passenger who did not have the steering wheel to steady himself. A carpenter built a wide seat for Mr. Chips that fit closely over the transmission.

Clarice and I reached agreements about who would drive, for how long, on this estimated three- to four-month-long trip. Each of us would sit two hours behind the wheel, whether or not we were driving a lot, little or not at all. We would try to gain mileage from daybreak to sunset and we would *never* race the motor (that's what kills automobile engines). We also agreed to carry along not more than two extra days' food provisions except for a stack of bananas, to light our Primus kerosene stove twice daily only, and to eat whatever the local population ate. This applied to Mr. Chips also.

Thanks to my years of experience in the war, I was confident that the WW II military Jeep was the most reliable motor vehicle ever made by man. Ours had to run, without visits to a service station, for 10,000 miles over the world's worst paths, tracks, washboard roads; in the terrible heat of the Arabian desert, or in the freezing storms in Afghanistan; in water, in snow or in powder dust; on gasoline which might be ten years old, full of condensation or dirt. Unless constantly retightened, Jeep springs would shake loose, bolts break off. I expected to be able to buy war-surplus major Jeep parts in all the larger cities of the world except in Afghanistan.

We purchased a complete set of wrenches and sockets at Bangkok's legitimate thieves' market. I anticipated the need for much tire patch glue and bought a whole sheet of tire rubber. One of our main problems would be to see that we never ran out of gasoline. I would have to top off the eight jerry cans that we had bought for gasoline and the Jeep's own tank whenever I found gas somewhere. I was to be also the poor fellow who would have to siphon out the gasoline to initiate automatic flow through the 20-foot-long rubber hose from jerry cans into the lower Jeep tank—without drinking too much of that awful stuff.

We practiced loading the Jeep and were shocked to find how little space there was in *Rosinante.* (A WW II Jeep may look reasonably big, but isn't.) The ten 5-gallon jerry cans took up a lot of the space in the back. Each of us had one aluminum Halliburton suitcase for summer and winter clothes; I had one box for tools, snow chains and Jeep parts. For Chipsy we needed one dish for food and one for water. Clarice and I would ride in *Rosinante* sitting on top of two woolen Army blankets and one down-filled war-surplus sleeping bag. The two Navy folding cots with mosquito netting we had brought from the States would be wrapped in our tarpaulin and tied across the top of the Jeep's hood. Two used spare-tire casings would be stuck between front bumper and radiator grille; a complete new spare wheel was mounted on its regular spare-wheel bracket at the Jeep's rear. Tow ropes, gasoline funnel and the long rubber hose to transfer the gasoline would have to be stored handily. We knew that we had to drive a greatly overloaded and crowded Jeep. If one of the U.S. Army's weapons carriers (three times the size of a Jeep) had been available in Bangkok, I would have traded in *Rosinante . . .* but most likely would have never been able to complete the trip because of its size and weight. Everything had to have its precise place and to be tied down so firmly that stepping on the brakes or hitting a deep rut in our path would not get us buried underneath our forward-sliding cargo.

Clarice and I looked at the National Geographic map, which showed no details of our intended route. We laid out our way to drive through major cities such as Rangoon, Mandalay, Agra, New

Delhi, Kabul, Teheran, Baghdad, Jerusalem . . . and famous sightseeing places or historical points like the Kohima Pass near the Burma border, Agra's world-famous mosque the Taj Mahal, the ancient fortress of Peshawar in India's North-West Frontier Province, the Khyber Pass (entrance into "forbidden" Afghanistan), the Hanging Gardens of Babylon near Baghdad, the Blue Mosque in Jerusalem and, perhaps, the Sphinx and the Pyramids. . . . Much depended on our and *Rosinante*'s health, the weather, our progress, adequate gasoline supply, cost of the trip . . . and if we were still talking to each other! We were in no hurry, had no timetable to meet, but wanted to get through and out of Afghanistan by mid-December.

Our medical kit consisted of Band-aids, one iodine bottle and three small boxes of aspirins. The U.S. Ambassador in Bangkok urged us to take along a .45 Colt automatic which he would lend us, but I knew too much about the Orient to accept his suggestion. If, by any miracle, I hit, disabled or killed one possible attacker, or two or even six with the six bullets in the gun clip, the three of us would be dead ducks within minutes. Orientals would never attack us unless they were in a large majority. Instead of relying on a gun, we decided to depend on Mr. Chips. We bought two whistles, one for each of us hung around our necks; we had fun training Chipsy to dash in the direction from which he heard the whistle's sound (in case we got separated in the jungle). Finally, I mounted a short flagstaff and an 8-by-12-inch Old Glory given us by our ambassador on *Rosinante*'s right front fender. Clarice's ability to speak French, Italian, Spanish and some Russian—besides English, of course—and my German and adequate Chinese ought to get us by in just about any part of the world. We simply would have to *talk* us out of any people problem. Finally—we carried U.S. passports, which in 1947 produced a magical effect in nearly every part of the world.

Having finished the essential preparations, we went sightseeing in downtown Bangkok; visited *wats* (temples) and the King's Palace; hired a boat which rowed us through the famed *klongs,* Bangkok's floating market. We even found an ice cream plant, which sold us their minimum quantity: one gallon of delicious vanilla ice cream! The three of us sat in the shade of trees on the lawn of the palace grounds and ate ice cream until we could eat no more.

13

10,000 Miles Across Asia by Jeep

At long last, we started the 10,000-mile drive due west. Our travel checks for five hundred dollars were safely tucked away. I had kept from the war my little plastic round-edged OSS escape compass (which could be swallowed in an emergency and found in one's body waste twenty-four hours later), but there was no need for the compass at this point. The first part of the way, from Bangkok to Rangoon in Burma, was due west and ran parallel to an abandoned single-track railway. A few miles out of Bangkok we got our first taste of what lay ahead for the next few weeks: an incredibly slippery road, unpaved and of red clay which stuck an inch thick to tires and shoes (similar to the surface I had encountered on the Burma Road in 1940). We met an occasional bus jammed with natives going to the market in Bangkok. Some people rode on its rooftop, holding on precariously to a handrail.

Our fuel tank was full; the Willys engine purred smoothly. We rolled and slid along at 10 miles per hour, in four-wheel drive, through rickety villages and past well-maintained red-and-gold painted *wats.* People smiled at us with black-stained betelnut teeth (if they had any left). After two hours we had our first of many, many flat tires—this one caused by a long rusty nail. It had punctured tire and tube of the left front wheel. Flats became so frequent as the mileage piled up that the tire problem could well have become the reason why we might have given up driving home. Since we were not blocking any mass traffic (an occasional bus or water buffalo led by a small, barefoot child coming out of nowhere and going what seemed to be nowhere, or a working elephant pulling a teakwood log chained to his hind leg, was about all that moved along this "highway"), we decided to fix the flat tire on the spot. The spare wheel remained in reserve. It was a tough operation merely to remove the defective wheel even before I tackled taking off its tire! Neither that first day, nor rarely afterward, did we have a sturdy enough base on which to place the truck jack and raise the axle of the fully loaded, very heavy vehicle. After loosening the five wheel nuts a few turns, I tried to jack up the Jeep chassis high enough to get the wheel off the ground to remove it. But the unpaved road and heavy Jeep made the tall jack sink deeper and deeper into the ground rather than lift the Jeep's wheel into the air, until, finally, the jack found a firm base several inches below the surface of the road.

Our World War II inner tubes were of poor-quality, newly developed synthetic material that stretched steadily, becoming larger and larger as the tubes got hot from supporting the heavy Jeep. When at long last I freed the inner tube, we had no basin of water into which I could submerge the blown-up tube and watch for air bubbles to find the place of the leak. Patching the inner tubes was not difficult when the trip began but steadily worsened as the tubes grew in all directions. The tube size increased 30 percent in the months to come; instead of nice little puncture holes, foot-long circumferential cracks developed. Once the hole or slot was fixed, inner tube and tire casing had to be re-joined between both halves of the wheel without pinching the stretched inner tube, a near-impossible task. All flats had to pumped by hand. No air compressor was here! Each time, I counted two hundred full pump strokes to fill the tire. It took one to two hours to fix one flat. Later on, I decided to leave tire repairs for the evening while Clarice was cooking the one-pot dinner, and to continue to drive using our spare.

Late the second afternoon, we reached a jungle village which, according to our odometer reading, was Kanchanaburi on the River Mae. Natives stared at us and Chipsy, then pointed in the direction of a prosperous-looking thatched-roof compound. There we found a German-speaking Siamese who invited us to spend the night in his place. A good-sized fire had been lit in the middle of his compound, to scare away wild animals. Our host was the son of the Thai Ambassador to prewar Berlin. He was in charge of the nearby Allied war cemetery. Its graves were those of former inmates of a Japanese prison camp in which thousands of British and Australian soldiers were kept under most inhumane conditions; over three thousand of them perished. The prisoners had been forced to build a railroad bridge across the River Mae, but the bridge was sabotaged by these same prisoners shortly after it was put into service. Had it still been standing we would have driven over it toward Rangoon. (By now you have probably guessed it: This tragic but beautiful place became the location for the famed movie *Bridge on the River Kwai.*) We found the prisoners' cemetery well kept, with freshly painted white crosses and flowers over each grave maintained by pretty young Siamese girls. (When I revisited this site during my trip to the Vietnamese war in 1967, the wooden crosses had been removed; instead, a large bronze plate indicating what was once here had replaced them.)

Our host knew of no way for the Jeep and us to get across the wide and very swift-flowing River Mae nor how to get to Rangoon any other way. He suggested that we return to Bangkok and start anew—heading north instead of west, all the way through the length of Thailand. We would enter Burma near the start of the original Burma Road into China; then head due west to Mandalay, leaving Rangoon to the south of us and continuing from there north to India. This man sounded knowledgeable and we followed his advice. We returned to Bangkok and our Hotel Europe, had a few good meals, filled up our gas tanks, bought more tire patches and two sections of a heavy WW II aircraft landing strip made of iron, to be placed underneath the Jeep wheels before entering a muddy stretch. (This turned out to be a stupid idea of mine: What we needed were *four* mats—two for the Jeep to sit on, the other two to place in front of its wheels! But the weight of four sections would have been too much for us to handle, nor did we know where to stow the heavy steel mats on *Rosinante.* Even the two mats which we had became a nuisance since I had to mount them on the side of the Jeep, blocking completely the entrance to one of the two seats.) We visited the ministry of communication to get the latest "road" information, updated our Thai exit visas, then headed north toward Ayuthia, the former capital. From there on to Takli, Lampang and Chiang Rai . . .

Wherever we drove or walked in Thailand, Chipsy was stared at as if he were a calf from the moon. Yet everywhere our American flag on the right fender caused admiration and respect. It took us two full weeks to traverse Thailand, a distance of about five hundred miles. Only in the immediate vicinity of towns were there any paved roads; these faded into the jungle within a mile after they passed through town, disappearing into dense, three- to four-foot-high grass, mud beds, or across streams without bridges. We learned to live in the jungle, how to remember our trail in case we had to turn back (it happened a few times), how to act when meeting people who were frightened to death when they saw Mr. Chips—although they were unafraid of an elephant! The Thai were the nicest and most helpful people we met on this journey; yet we could communicate by sign language only. Wherever we went, people stretched out their hands and pressed their palms together, begged us to stay with them as their guests in teakwood bungalows (built on 12-foot-high stilts to keep snakes and wild animals away), insisted on feeding us and absolutely refused to accept any payment. They brought us bananas, other fruit and vegetables; always accompanied by big smiles and signs of sincere friendship. Once, a village butcher invited us to select any piece of pork from a pig he had just slain. It had been sliced right down the middle and hung upside down from four stakes driven into the ground in the village square.

Thai Buddhist monks, shaven-headed and in deep-orange wraparound robes, chopped down one side of the wooden railings of a long, narrow bridge leading into their compound when the right and left railings were a few inches too narrow to accommodate our packed Jeep. Natives materialized out of the jungle when they heard *Rosinante* groan and strain as we were stuck deep in mud or in a shallow riverbed. After an initial shy encounter (at first women hid their children behind their long skirts when they saw us) all people—young and old, men or women—tried to be helpful to push, pull, lift and float the Jeep across mudholes, jungle growth or paths blocked by fallen trees. Although they had plenty of opportunity to steal something from us, no one ever did. If we did not spend the night with Thais sleeping on the floor in their log houses high off the ground, we camped under our tarpaulin tied to tree branches and to the Jeep's side, with Chipsy on a long leash. We slept on our Navy cots under mosquito nets (Clarice insisted on using white bedsheets and pillowcases). We heard wild animals breaking through the underbrush nearby; they were probably as scared and nervous about the strange sounds and smells coming from us and Chipsy as we were of them.

Admittedly, there were moments when we wished we had never started that trip—especially when we saw a large Lockheed Constellation passenger plane fly smoothly high overhead and imagined its air-conditioned interior, soft seats, service by hostesses . . . while we were stuck in a mudhole. But the challenge to do something which, to the best of our knowledge, nobody else had ever even tried to undertake with a single vehicle got the better of us and we struggled on. (I had heard that, years earlier, a British Rolls-Royce factory convoy consisting of three cars and one supply truck had driven from London to Rangoon.) Twice, two-tenths of a mile in one day was all that we managed to progress when natives assembled a raft out of bamboo poles to float the Jeep across a wide stream. When there was no one to help in pushing *Rosinante* out of a rut, or when she had sunk into mud up to her chassis, we had to unload everything from our Jeep to make her lighter, carry suitcases and gas cans hundreds of feet forward till we reached solid ground. Clarice or I then drove the empty vehicle in four-wheel drive very slowly while the other one pushed it, only to drop into another mud pool a few hundred feet farther. . . .

In the north of Thailand we came out of the jungle and passed a small city, Lampang, near the end of the single-track north–south railroad traversing the country. One freight train rolled daily on the track. It frequently stopped to let farmers and their produce on and off. We had just bought on the open market (where women sat on their haunches, selling their neatly arranged produce) half a dozen chicken eggs, which Clarice carried on an open banana leaf, with Chipsy's leash looped around her wrist. Suddenly, a train conductor blew a whistle similar to our own emergency whistle. Chipsy took off like a bolt of lightning! The eggs fell off the open banana leaf, and Clarice stumbled. . . .

When we had about ten miles left to go in densely wooded Thailand before we crossed into Burma, we were led by natives to a well-built British-style bungalow surrounded by tall trees; it was two stories high and had a screened teakwood verandah around the four sides of the upper floor. There, by George, we found a young British chap living with his servants. He was the manager of a specific district of the British-American Tobacco Company, on a two-year assignment. This young man had come from London and was, of course, delighted to see other European or American faces. He couldn't do enough for us. His servants prepared a hot bath for Clarice. A candlelit dinner was served, each of us having a silver finger bowl with rose petals floating on water next to our plates. He had his own electricity-generating power unit, which made it possible for us to hear news from the BBC over his shortwave radio, have electric light in the guest room and even an electric cooling fan. It just didn't seem possible that we were dining and resting in the midst of jungle and tobacco fields between Thailand and Burma! That one night we slept in real beds. The young Englishman urged us to stay with him for a while and let an organized group of bandits pass (they were expected to rob the annual silk-merchant caravan traveling from Thailand to Burma. "This is a tradition," he told us). But we decided to keep on moving since our progress had been slower than I had figured. My fear of a snowy winter in Afghanistan was the key factor.

We didn't quite believe the tobacco man's report on the bandits (the plantation manager was drinking something stronger than tea when we were with him)—until we ran smack into them the following evening just as it was getting dark, a few miles inside Burma. There was nothing we could do at that point but play it cool; we tied our tarpaulin between the Jeep and a couple of trees next to their tent camp, which was already set up. There must have been about one hundred people, including their womenfolk decked out in silver jewelry. The bandits wore parts of British uniforms and Eisenhower jackets. They carried American rifles and sat around a campfire, their small horses grazing in the dark. Their cook was Chinese. The man who invited us to share their delicious chicken dinner seemed to be their leader. He spoke a little English; our conversation was friendly but uneasy. By nine o'clock we wished them well, went to bed in our sleeping bags on cots, hoped to wake up in the morning with *Rosinante* still next to us and neither of our throats cut. Well, we awoke in the morning, found the bandit camp broken up and the people gone—silently, and not a bit of garbage left behind! A large cake had been placed for us on the hood of our Jeep. Why Chipsy did not bark we never could figure out. Perhaps he was just as tuckered out as we were?

Fixing flat tires became the most unpleasant and exhausting routine, particularly since the glue of the older patches became brittle when the inner tubes became very hot because of overload and road conditions. This brittleness of the patches caused extra leaks. In addition to the problem of finding the leaks and the hard job of pumping each tire by hand in exhausting heat and humidity, a totally unexpected problem arose: The 18-inch-long, small rubber hose from tire pump to valve stem of the repaired inner tube began to split at each of its ends when the pressure was at its highest during the final strokes. Our tiny pump hose had lived through World War II and was now brittle. Hose cracks called for cutting off the split end squarely, thus shortening the rubber. That's what I did, crack after crack, until the hose connecting our pump to the tire valve got so short that there was not enough length left to place the pump on the ground next to the tire. Instead, the pump had to rest on the wheel itself, four to five inches away from the valve stem! Fortunately, the last bit of hose lasted, with Clarice pinching both rubber ends with two pairs of pliers and after I had wrapped all the Band-Aids we had left around its dried-out rubber. (We got a new pump in India, a present from the U.S. consul there.)

To even out the load on each of the four wheels and keep the tire temperature from climbing even higher, we had driven *Rosinante* continuously in four-wheel drive—until Chipsy stood up shivering instead of sitting on the board above the transmission. I gave him a few pats on his behind and told him to sit down because he was restricting free arm movements of the driver. So the poor dog crouched with his knees bent, trembling. . . . Only then did I see the faint smoke rise from under his seat: The transmission had overheated! We shifted back to the standard two-wheel drive—and had the anticipated increase in rear tire failure rate. As I had expected, at Mandalay we found many discarded military Jeeps from the war, but none looked confidence-inspiring. So I added some transmission oil, and we kept on driving with a "singing" gearbox between us. We added a mat for Chipsy's comfort until we could replace our transmission with a good one, two thousand miles later.

In many sections of Burma we drove on rough segments of a steep mountain road built by the Japanese military in 1942 and 1943. The driving progress was slow. One of us would shift the Jeep into four-wheel drive, in "low low" gear at 1¼ miles per hour speed, while the other and Chipsy would walk ahead of *Rosinante* to stretch their legs, and then wait. . . . We never ran the engine fast; I retightened the bolts and nuts of the front and rear suspension system every night. Such taking care of our equipment paid off in an amazing reliability.

During a few days' stay at Mandalay (Kipling's city was totally destroyed in British-Japanese fighting in 1945), we got fresh food and some sleep in a patched-up school auditorium which we shared with American and German missionaries who had come down from "the hills." We left Mandalay in the direction of the Brahmaputra River beyond the northern frontier of Burma. All bridges over the wide Irrawady and Chindwin rivers had been blown up by the retreating Japanese and had not yet been rebuilt. We talked with the few remaining British officials who were to transfer the government to the Burmese. They offered us—without charge—a barge for *Rosinante* to be towed by a tug; first downstream the Irrawady, then against the strong current of the Chindwin, to reach the beginning of the long, curvy mountain road to Imphal near the border of India. This sixty-mile-long boat ride took six days, gave me a chance to do a thorough maintenance job on *Rosinante,* change the engine oil, lubricate everything, and even cut and nail a pair of heels on Clarice's shoes. The Burmese boat crew fed us fish they caught; during nightly stops at villages, we got rice, eggs and fruit. North of Mandalay we found a matriarchal society: Women did the heavy manual work, ran the stores, smoked cigars while their hubbies stayed home, took care of the children, did the cooking and took it easy!

Burma was under British rule for a few more months, as we were driving through it. Guesthouses along the roadside, vacant and without any furniture except bed frames, dusty and not too clean, became more frequent. (Cords across the wide bed frames served as mattresses.) These wooden, single-floor motels, called *daks,* were owned by the government and provided shelter to officials who traveled across the country. We just pulled up in front of one of these three-room *daks.* Without exception, they were totally unoccupied. There were no locks on any doors and no windowpanes, no kitchen, no light, no toilet. But we had a roof over our heads. We got out of our sleeping bags half an hour before sunrise, and went "to bed" immediately after dinner and after the last flat tire was fixed. Usually, a hand-operated water pump was near the site where we could wash and I could shave.

The mountain pass beyond Imphal, the Kohima Pass, was still littered with wrecks of Japanese light tanks and British artillery. A sign nailed to a tree stump at the zenith of the pass read: AT KOHIMA—IN 1944—THE JAPANESE INVASION OF INDIA WAS HALTED. A vicious battle had taken place here, three years earlier. Exhausted Japanese Army units, supported by several Indian regiments who had switched their allegiance, fought against the British —and loyal Indian troops. The advancing Japanese were preceded by a group of Indian women who were forced, or volunteered, to protect the Japanese from point-blank gunfire of the British. We could tell from branchless tree stumps that the battle must have been furious. It had lasted over a week before the Japanese ran out of ammunition. Their long supply lines from Rangoon to Kohima had frequently been cut by British and American glider troops. Commanded by the brilliant British General Orde Wingate, these troops had landed behind the Japanese lines. His staff included famous American officers such as John Alison (who had flown "my" Zero on its first flight in China in 1942).

Mile by mile, day by day, flat tire by flat tire, we moved first northwest and then due west. We traveled past a sand-filled gasoline barrel with a sign, BORDER CONTROL INDIA, stuck into it; a stamp was pressed into our Jeep triptyque and U.S. travel passports. We were informed that gasoline had been rationed for several weeks in India; it was available at the few hand gas pumps with their glassy cylinders in exchange for ration coupons. These were made available by the district manager, in his office. (We, of course, had not the slightest idea where and what an Indian district was.) The price was thirty cents for an imperial gallon of gasoline, which is about 20 percent larger than an American gallon. Our path led us through giant tea plantations to the mile-wide Brahmaputra River, which we crossed near Gauhati on a large barge pushed by a teakwood-fired tug. We continued north to Siliguri on asphalt roads. There we had a spectacular view of one of the world's highest mountains, 28,146-foot-high Mount Kangchenjunga, and the tip of Mount Everest. Within sixty miles of Tibet we swung around to the southwest for a few hundred miles of fair road until we came to the Ganges River, on which human bodies floated downstream just as I had seen in 1939.

There was no road bridge across the wide Ganges, only a one-and-a-half-mile-long, two-track railroad bridge connecting both banks. An Indian Army officer, on "our" side, refused us permission to cross over the railroad ties. He wanted us to load the Jeep on a railroad flatcar, then cross the bridge. All this would have taken at least one full day. But after our persuasive efforts, backed by showing him George C. Marshall's signature on the "issued by" page in our passports, the officer exclaimed, "If General Marshall signed your passports, that's enough for me to halt the train traffic!" After we were certain that traffic in both directions *was* halted, we began the crossing. The officer was under the impression (and we left him thinking so) that Clarice and I were on an official U.S. Government mission. How come? Our passports carried the printed but real-looking signature of George C. Marshall, the then Secretary of State (as did every other U.S. passport issued at that time). We also casually displayed a sealed envelope from the U.S. Embassy, Bangkok, to our embassy, New Delhi (it was from a girl secretary to her friend stationed in India). Both Clarice and I wore military khaki clothes, which also might have added to the officer's impression that we were on official business. The Jeep and we three took an awful beating hobbling over the thousands of railroad ties, but we made it to the other shore and thanked the Indian officers there.

We enjoyed a few hundred more miles on very good, tree-shaded roads, with only an occasional truck and a great number of camels passing us. One little village we drove through hosted masses of two- to three-foot-tall monkeys; one of them picked a fight with Chipsy, grabbed him by the nose and pulled out some hair (which never grew back). These agile animals lived on corrugated rooftops, in trees, on the road—totally free.

At long last we arrived in the middle of northern India. Clarice had the chance to visit the incredibly beautiful Taj Mahal at Agra.

To me, the Taj is the most beautiful building in the world. From there it was only 140 miles to India's capital, New Delhi; we visited our American embassy and were delighted by a generous gift of fifty gallons of gasoline.

In the junked-car market in Old Delhi, next to New Delhi, I looked for a used Jeep transmission. Ours was just about done with. I bought a complete secondhand replacement for ten U.S. dollars. There was no way of telling if it was any good. A hard two-day job for one man, i.e., me alone, to remove and replace the heavy transmission with its front wheel drive, right in front of our hotel in New Delhi, was not in vain. A test drive showed that I had, by chance, bought a good transmission.

A reporter from the United Press who did not believe that we had come by road from Bangkok interviewed us. He warned us of the dangers of the civil war which had just begun between Hindus and Moslems 250 miles northwest from Delhi, exactly where we were headed! His newspaper article describing the Neumann family and their Jeep trip to India showed up in many U.S. newspapers, and was the first word Clarice's parents had had about their daughter in a long time. Plans for a partition of India into Pakistan and Hindustan were being translated into reality at Lahore and Amritsar, on the road to the ancient fortress of Peshawar, through which we planned to drive en route to Afghanistan. Two million refugees, Hindus moving east, Moslems west, clogged the road. Killed Indians lined both sides of the highway near Lahore. An American official in New Delhi, too, had warned us that it would be impossible to cross the battle lines already drawn, that troops of both sides were in their trenches and that it would be highly dangerous to try entering Pakistan from India. Even U.S. officials had been unable to cross from one self-proclaimed country into the other, he said. We went ahead anyhow, found their report true, except that Clarice, Mr. Chips and I *were* able to cross the trenches. The Americans did not use my secret weapon: the camera.

At the barbed wire barricades I offered to take photos of the officers and their troops. Both parties accepted. During a fifteen-minute truce, barbed wire roadblocks were moved aside, and soldiers and their officers emerged from their trenches in full battle gear. Just for the fun of it, I persuaded both sides to pose once together: Hindustani and Pakistani soldiers stood at rigid attention, each in their distinctive uniforms and with their own standards. We thanked both the Hindus and Pakistanis, then bumped over the broken-up road and headed for Afghanistan via Peshawar and the famed, winding Khyber Pass, the only entry from the east into that rugged country. We ran into the annual colorful exodus of over one thousand Afghan nomads leading their sheep and camels, with children riding atop, chickens tied to the camels' saddles. Their leader spoke some English. He told us that he had three wives (with him!) and sixteen sons, but really didn't know how many daughters! As in the old China, first marriages were prearranged by parents. Some selections did not turn out too well. Since the young bridegroom could usually not see his wife-to-be until the wedding ceremony was over, the surprise was often great. Once a young wife lifted her scarf for the first time to face her new husband and asked him to whom else she might show her face. Her husband's answer: "I really don't care—as long as you don't show it to me anymore!" Higher and higher, colder and colder went the winding pass. A partially completed railway and tunnels built by Germany before the war went alongside the pass. The first snow showed up on the unpaved road as we approached Kabul, Afghanistan's capital. We found a couple of Morrison-Knudsen camps, a construction company in Boise, Idaho, en route, where we were given shelter and food.

The only asphalt road in all of Afghanistan in 1947 stretched in front of the king's palace. Soldiers in German uniforms stood guard there. King Amanullah, an admirer of Germany, had reigned here from 1926 until he was forced to abdicate, in 1929, because he had tried to move his backward Moslem country too fast into the twentieth century. For example, he had forbidden women to remain in purdah any longer (i.e., no veil to hind behind!).

We spent a day with the American military attaché in Kabul to get a good night's sleep in his home and local know-how on road and weather conditions ahead. We bought fresh food, were given all the gasoline we needed, loaded *Rosinante* with the Afghan staples of raisins and nuts, and received directions to the next Morrison-Knudsen camps which we planned to visit en route. We left Kabul, swinging south around the huge mountains, without marked roads or any direction signs. The snow on the ground was frozen hard, the country totally barren. We spotted an occasional telephone pole without any wire and ancient mud fortresses, which confirmed our general direction. Herds of thick-wooled sheep were moving to the south. I bartered for a sheep's wool jacket for Clarice, who was suffering from a typical case of malaria. (She had probably been infected by mosquitoes in Thailand two months before; the malaria now broke out in the freezing weather.) There was no doctor for a thousand miles, nor had we any quinine or Atabrine tablets. Instead, I melted snow three times a day to make for her cups of Lipton's chicken noodle soup, of which we had bought two dozen small paper bags in Hong Kong. I spoon-fed the soup into Clarice's mouth and hoped for the best. Uncomplaining, Clarice rode bundled in her sleeping bag, which I tied to her seat with the Jeep's safety belt; Chipsy lay half on her lap, sharing his body warmth with her.

Because I was not sure of the protective strength of the antifreeze mixture in our radiator, we could not afford to stop the Jeep's engine during one very cold night and had to let it idle fast. During that night we slept in an Afghan guesthouse (ten mud rooms without *any* furniture, lights, bathrooms or water to drink, but a shelter against the icy wind). We drove through Kandahar's main street with its one-level mud houses and open stores. Only bundled-up men were visible. Two hundred miles farther was the American winter camp of Morrison-Knudsen, to which we delivered mail addressed in care of the U.S. Embassy in Kabul. These American surveyors and construction folks, some accompanied by their wives and children, had bundled up in five prefabricated and winterized buildings. They hibernated there for four months. We were welcomed warmly especially since we had mail for them. That night, after an American dinner party they held for us, all three Neumanns (Mr. Chips included) got sick from overeating! Before waving us farewell, the manager filled our tanks with gasoline and gave us a much-appreciated roll of tire patch material.

Five days later we drove into the town of Herat at Afghanistan's western frontier, high above Iran. The weather became a bit less frigid, and there were brown spots showing through the snow cover. Clarice's malaria had abated although she still got some shivers every thirty-six hours. *Rosinante* continued to behave beautifully. We had had no flat tires for the last 1,800 miles, thank the Lord! My hands were so cold and stiff that I wonder if I would have been able to do any tire repair work. In a climate like this, we wore *all* the clothes we had and both pairs of gloves; the Jeep's sides were pulled down tightly, the exhaust pipe bent to heat the inside of our knees . . . still it was *cold.* We made 100 to 150 miles per day, driving during the nine hours of daylight, and slept with native Afghans, their goats and sheep in mud beehives that looked like Eskimo igloos. The mere idea of sunshine and warmth we expected to find "beyond the western horizon" made us press on as fast as we safely could. During the fifteen-mile drive downhill from Afghanistan into Iran across no-man's-land, into the setting sun, we encountered an ancient Harley-Davidson motorcycle with sidecar pulled by a horse which was led by a man! The owner, a beturbaned driver, was riding his powerless cycle at two miles per hour.

It was dark when we reached the much lower frontier control post of Iran. Soldiers with fixed bayonets ran toward us in the glare of our headlights. The guards yanked at our rolled-down side canvas but backed up in a hurry when Chipsy let out a nasty growl. We had spotted from the high Afghan plateau the dim lights of the little town of Turbat-i-Jam, twenty-five miles into the valley on the Iranian side, and decided to spend the night there. Word of our visit spread like wildfire. The population of the sleepy town was excited over a Jeep visit by two Americans and their dog. The mayor invited us to stay as his guests, then called in his friends, who had already gone to bed. With great ceremony he brought in a dish with a bar of Ivory soap and presented it to us as he said with a flourish, "Palmolive!" He spoke halting French. A hospitable party lasted late into the night. All three of us slept in the mayor's dining room on piles of Persian rugs. The local schoolteacher asked us to visit his class, but we had no time to spare: Snow was expected to fall any day—and once again was the deciding factor. Somehow I had hoped that all of Iran would be a warm country, but we were disappointed. Next morning, we drove on the packed sand road leaving Turbat-i-Jam for the northwest, to the Russian-Iranian frontier town of Meshed, 140 miles away. The Russian influence was clearly visible: Men were bundled in heavy coats and blankets, and sat on low benches in front of their stores, smoking water pipes and drinking tea heated in samovars over open fires; we even filled up our tanks at a real gas station! Five hundred miles farther west, over frozen but snowless washboard roads, lay Teheran.

14

"From Teheran on It's Easy . . ."?

It took us three days to reach Teheran, descending into this large city from the bare, brown hills surrounding it. We had reached a landmark: From Teheran on, there were paved roads to the west, north and south. These roads had been built by American engineering troops for the fast supply of weapons and vehicles to our Russian ally during World War II. It was the first time since we had left New Delhi that we stayed in a real hotel and enjoyed the luxury of a bathtub with hot water (even if there was only one bathroom for two floors). We thawed, peeled off the frozen dirt from our hands, washed worn-out clothes, ate in a restaurant. What luxury! We wrote Clarice's family that we were still alive, shopped in stores and real Mideast bazaars with their low curved ceilings; tea was served in small glass cups and silver holders, by little boys . . . whether or not you were a customer. Some of the women we saw were totally veiled, as they were in other Moslem countries we had driven through.

Our U.S. consulate in Teheran tried very hard to help us obtain the next visas for Iraq, Syria and Lebanon. At that time America's political standing was so bad in the Mid East (as we had been warned by General Chennault it would be) that in spite of polite receptions by Arab consular officers, we were told that the Arab

League in Teheran had decided not to issue any more visas to Americans for the time being. This included transit visas. A severe crimp in our travel plans . . . for a few minutes anyhow. We decided to bypass the explosive Middle East and secured Turkish transit visas. We changed our planned route: We would now drive north to the Iranian-Turkish border, cross the mountain passes on the way to Erzurum and head for Turkey's capital, Ankara, and famed Istanbul with its Golden Horn.

Sitting by a comfortable radiator in Teheran's Park Hotel, eating dates and figs, with Mr. Chips at our feet, Clarice and I discussed what we should do upon our return to the USA. Christmas of 1947 was just around the corner. By this time it had become clear to both of us that we would have had enough of driving *Rosinante* once we reached the Mediterranean Sea. We decided not to go back to California, although I had liked the West Coast life style very much and had enjoyed my job, the colleagues at Douglas Aircraft and the West Coast weather. Clarice, however, had experienced mainly foggy or rainy days during the three winter months she spent in our rented room near the Santa Monica beach. She urged me to find a job on the East Coast. Most large airplane plants were in the west; but the aircraft engine companies General Electric, Westinghouse, Pratt & Whitney and Curtiss-Wright were located in the east. Clarice was aware that GE had a plant in Bridgeport, Connecticut, making small home appliances there. So I wrote an airmail letter from Teheran to GE in Bridgeport asking them to transmit my request for an interview to wherever they were making jet engines. There was no hurry to reply, I wrote, since I would probably be "fixing flat tires for another three months" before I would be available for work in the States. I gave GE a forwarding address.

It was snowing heavily when we left the warm and cozy hotel life in Teheran. We bundled up again, cautiously purchased with our meager funds canned food, raisins and nuts, and headed north toward the Turkish frontier via the only major Iranian town en route, Tabriz. Ice and snow built up on our windshield, which had neither a defroster nor an automatic wiper. Except for an Iranian Army truck heading in the same direction, we saw no other vehicle on the well-marked highway covered with frozen snow. In spite of chains on all four wheels, *Rosinante* got stuck several times in two-foot-deep snowdrifts. The Army truck following us was loaded with twenty friendly Iranian soldiers in warm winter uniforms who were armed with snow shovels. They drove more slowly than they could have, I am sure, to help us dig out the Jeep.

It was already dark when, late on Christmas Eve 1947, we arrived at Tabriz, less than one hundred miles away from the Turkish border. A policeman rode on *Rosinante*'s hood to point out where we could find a hotel and also the American consulate. A small, warm guesthouse spread out the welcome mat for the three of us. We dropped by the official U.S. residence that same evening, and found two junior American consular officials debating what parts of an electric toy train they should order, to have something to do in this Godforsaken post. . . . They told us what we were afraid to hear: The only road leading north into Turkey and Ankara was officially closed because of twenty feet of snowdrifts! What now? The choice was among hibernating in Tabriz for four months, or driving south to the Persian Gulf across all of Iran, or heading due west over a high mountain range to Iraq's oil fields at Mosul. We also learned from the Americans that there was an Iraqi consulate in Tabriz. Why not try next day and see if the Iraqi consul possibly had not yet heard, in this remote part of the country, of an Arab League ban for American travelers on obtaining transit visas through his country?

We took a chance: To avoid any military appearance, we left *Rosinante* at the guesthouse and hired a droshky, a Russian-style horse-drawn taxi, to get us to the Iraqi consulate. It was Christmas Day—December 25, 1947. The Iraqi consul opened the door himself. He was a most amiable man in Western dress, speaking French quite well—and obviously lonely. Conversing in French, Clarice told him of our trip so far, that the road to Turkey seemed impossible until spring because of the snow, and asked him for his advice. Although he recognized that this would be more than one thousand miles out of our way, the flattered consul urged us to go to Europe via Iraq. When we replied that as Americans we wouldn't feel safe in his country, he got up-tight and *insisted* on giving us, gratis, the Iraqi transit visas to Baghdad via the oil fields at Mosul, two hundred miles west of Tabriz. Reluctantly(!) we accepted his offer. While he had the two visas prepared, the consul served us delicious sweets. I played customer chess with him for four long hours. (We, of course, died a hundred deaths waiting for our stamped passports.

We were afraid that perhaps his office would phone his superior in Teheran.) After getting our visas, we went sightseeing in a movie-style dim bazaar divided into a network of stalls, row upon row illuminated by oil lamps. The beturbaned salespeople were sitting shoeless on piles of Persian rugs, peddling silverware, Persian clothes, food, bolts of woolen material. . . .

At sunrise next morning, we headed west to the southern shore of Lake Urmia, toward Mosul. The consul's unawareness of the latest Arab League decision not to issue any more transit visas to Americans was a real break for us. His ignorance was matched by his lack of knowledge on how to get from Tabriz to Mosul in two to three feet of snow. We hoped that whatever communication existed between Teheran and Tabriz would be slower than getting our Jeep across the Iraqi frontier! But for the fourth time on this journey, snow fouled us up: It was heavy and deep in the mountains south of Lake Urmia, without any indications where the road itself was supposed to be. After being briefed in Mahabad by the commanding general of this Iranian area, who also became our host for the night (he was proud of his good-looking tall wife and wondered how he could get her "to Hollywood"), we reluctantly returned to Tabriz. Such a decision was not made easily because the next step was a real gamble, the greatest we were to make on this 10,000-mile trip. Perhaps we took a bigger chance than we should have taken: In a thirty-six-hour *nonstop* drive, except for changes at the wheel, we drove back to within fifty miles of Teheran, then turned sharply on the only other road to the southwestern Iranian city of Hamadan, a total distance of 652 miles! Fortunately, there was heavy tanker traffic on the road during the day, packing down the snow. Nobody drove after dark.

We had made good progress during the first part of that marathon drive, grinding forward at 15 to 20 miles per hour. Shortly after midnight, however, when we were only eight hours away from Hamadan, suddenly our Jeep began to lose power. We faced the first pure engine problem since leaving Bangkok! Within a minute the engine came to a complete halt. An icy wind was howling, snow was falling heavily, and we were stuck in the midst of the deserted Zagros mountain range parallel to the border with Iraq—without power or heat, in pitch-darkness, at 8,000 feet altitude. There was not a single other vehicle on the road.

For our physical survival I simply had to try to get faithful *Rosinante* going again, before her engine and we froze fatally. I opened the hood and swung one of the headlights back (original Jeep headlights were mounted on a hinge to shine backward into the engine compartment). With frozen fingers I checked the fuel pump and gas line leading to the carburetor by disconnecting the pipe coupling, then cranking the engine by hand. Because fuel flowed out at the loosened connection, the trouble had to be *inside* the carburetor. Barely able to hold a screwdriver with my stiff fingers in the gusty, icy wind, I removed the cover of the carburetor fuel bowl, disconnected several tiny wire clips and placed them on the flat part of the Jeep fender. In the light of my weak flashlight, I spotted a tiny ball of ice inside the carburetor bowl, the result of frozen water condensation. Hoping against hope that this was indeed the cause of our problem, I removed the ice and began to reassemble the carburetor. To my horror, I could not find the two little wire clips: The wind had blown them off the fender, into the snow on the ground. There I was, past midnight, dead tired and shivering, trying to find in the soft snow two tiny wires the thickness of a hairpin, but much shorter. A desperate search with my flashlight helped me to find one of the two pieces; I had to make a second clip out of stripped electric wire I cut off from the Jeep's taillight. The moment of truth came: I sucked ice-cold gasoline out of the fuel line coming from the tank and spit it into the open carburetor air inlet, then reattached the line quickly. Clarice pulled the choke out. I gave *Rosinante* one yank with the engine crank, using the last bit of power I could muster. . . . The engine started! No words can describe the exhilarating feeling of accomplishment when the Jeep lurched forward again; its engine purred smoothly, the exhaust pipe began to warm up our legs, we stopped shivering, and soon were heading down the snowy mountain range toward warmer Iraq. Just lucky, I guess . . .

Why the great rush? We didn't want to give the Iraqi consul in Tabriz (who told us that his people had noted in our passports that we would cross the frontier into Iraq at Khaneh, near the Mosul oil fields) the opportunity to communicate with his boss in Teheran that he had issued two transit visas to Americans with a dog. This could have caused the Iraqi frontier guards to stop us. Whether he ever found out that his people—by pure luck for us—had mistakenly specified in our passports the much more southerly border crossing point into Iraq which we were now going to take, we'll never know. Obviously, the border guards—after an uneventful check of visas, of the invalid Siamese automobile triptych, of the wrong engine number and so on—had not heard what the Arab League had decided. They even offered us their guardhouse as *Nachtquartier.* We accepted and remained the night as the guests of the Iraqi military who previously had warned us against traveling at night: "Bandits will kill you and rob your Jeep!"

On December 31, 1947, we crossed the Tigris River outside Baghdad and checked into Hotel Semiramis. We had a drink at the bar in honor of the New Year but were too tired and worn out to join in the general celebration.

On the second day of 1948 we drove two hundred miles out of our way, looking for the Hanging Gardens of Babylon. We admired the "brickses" (as our volunteer guide, an ancient Arab who loved art, called them) which made up the famous bas-relief figures thousands of years old. Palm trees, sunshine, flowers . . . what a good start for the New Year! *Rosinante* practically flew across the bridge over the Euphrates River toward the Hashemite Kingdom of Jordan, eight hundred road miles away, for which we had obtained visas while in Baghdad. We also got Syrian and Lebanese visas—but did not try for a British transit visa through Palestine.

The road from here to the city of El Mafraq was a few hundred miles of a single-lane tar strip covered by blown-over sand. Poor *Rosinante!* Soft, warm sand dunes like those in the movie *Lawrence of Arabia* bordered the road as far as the eye could see. A large oil pipeline was being laid to Haifa in Palestine via El Mafraq by Aramco (Arabian American Company) workers in Arab checkered headgear. At that desert town we had to decide whether to drive north to Damascus in Syria or south to Amman, the capital of Transjordan. We had increased *Rosinante*'s cruising speed to 25 miles per hour—and felt as if we were flying! Still taking turns at the wheel every two hours, we pulled out the hand throttle and pushed the windshield forward. Bless the warm breeze blowing into our faces and the smoothness of the great Willys engine! We stopped at three oil pumping stations, were always invited to stay in the American manager's guesthouse amidst an oasis of palms, swimming pool and air-conditioned buildings. There was no more doubt: We were again in the Western world! The misery caused by nature was behind us; fighting our way through jungles in Siam and Burma, and the freezing weather in Afghanistan and Iran were years past, so it seemed. We felt like a million dollars, the three of us, having lost weight and having lived during the past few months about as simple a life as existed on this earth. Chipsy's meals still consisted of the same food we ate: fruit, bananas and even oranges. Whenever we stopped briefly to change drivers, he took off in search of better food, but returned at the blow of our whistle, jumped onto his seat on the transmission and must have thought, What next?

In Transjordan we followed a truckload of Arabs waving black flags and rifles in the air. They also carried signs: VOLUNTEERS FOR PALESTINE and DEATH TO THE JEWS. Outside of Amman, in plain view of the city, we were halted by a roadblock manned by soldiers of the Arab Legion, the well-trained, well-disciplined small Jordanian army equipped by the United Kingdom and commanded by the British general Glubb Pasha. Arab soldiers in red/white-checkered headdress carefully examined our visas (but not so carefully the Jeep's engine and chassis serial numbers). On three separate occasions we were asked if we were on the way to Palestine. "No. To the American consulate in Amman," we replied.

"Okay," they said, and let us pass.

In downtown Amman we inquired at the post office where we could find the U.S. consulate. To our great shock, we were told that America had never recognized Transjordan, thus did not have a consulate here. . . . What now? Word spreads fast in the Middle East; an unfriendly crowd had already gathered around the Jeep (its RAF insignia still visible on the engine hood, the American flag neatly folded in our suitcase). The Jordanians pushed close to *Rosinante;* a feeling of animosity was in the air. We made our way firmly but politely through the crowd and checked in at Hotel El Raschid, the only hotel at the center plaza of Amman. I even went to get a haircut, and was plenty worried when the barber picked up a 10-inch-long razor to shave my neck. Naturally, we were concerned about the Jeep and the things in it. The hotel manager persuaded us to hire a sit-in watchman. We did. Next morning, both he and our woolen blankets had disappeared.

Seventy-ninth Congress of the United States of America
At the Second Session

Begun and held at the City of Washington on Monday, the fourteenth day of January, one thousand nine hundred and forty-six

---

AN ACT

To provide for the naturalization of Master Sergeant Gerhard Neumann.

Be it enacted by the Senate and House of Representatives of the United States of America in Congress assembled, That upon compliance with all other provisions of section 701 or section 702 of the Nationality Act of 1940, as amended (56 Stat. 182–183; 8 U. S. C. 1001–1002), Master Sergeant Gerhard Neumann, Army of the United States, Army serial number 10500000, may be naturalized pursuant to either of said sections as may be applicable, notwithstanding the fact that at the time of his enlistment or induction into the armed forces of the United States, he had not been lawfully admitted to the United States and was not a resident thereof.

Speaker of the House of Representatives.

President of the Senate pro tempore.

APPROVED
JUN 15 1946

It required an Act of Congress before I could become a citizen of the United States. The bill was introduced in May 1945 and finally signed by President Truman in June 1946.

The roads in Thailand frequently led us to wonder why we had ever decided to drive from Bangkok to Palestine.

Clarice driving across a temporary bridge in Burma

Hindustani and Pakistani troops, after removing barbed-wire barricades for us

Mr. Chips sitting on Rosinante's hood, before we sold the Jeep in Jerusalem

Clarice at the Tel Aviv airport, with Mr. Chips in the "TWA dog flight-bag"—our own creation

"Full Speed or Bust" was my order for the first run of the GOL-1590 (VSXE) demonstrator jet engine, which used advanced technology from stem to stern in 1953.

The variable stator patent is owned by GE, but with my name as inventor. This novel feature improved engine performance and fuel consumption dramatically.

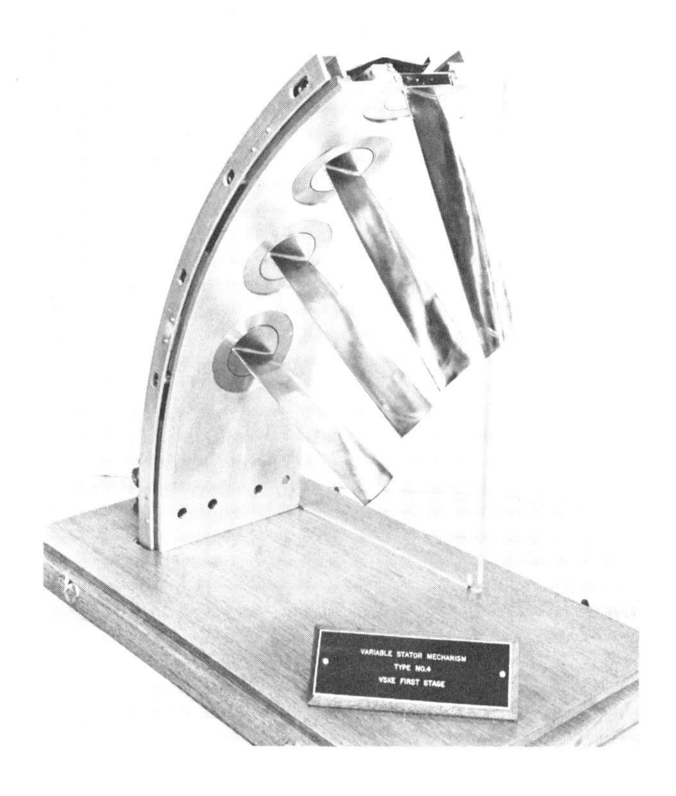

With Captain Bill McCurdy in the Lockheed F-104 "Starfighter," in which we made a supersonic flight below sea level

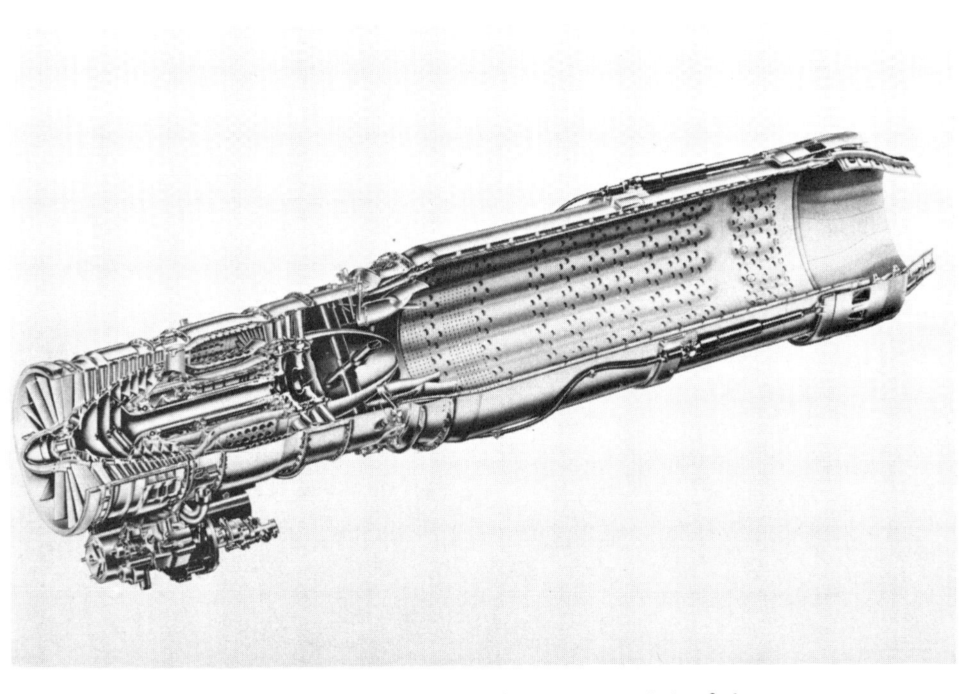

Straight jet engine *(above),* used on supersonic jet fighters and bombers, and the new-type GE "high-bypass" engine *(below)* used on modern wide-body transports

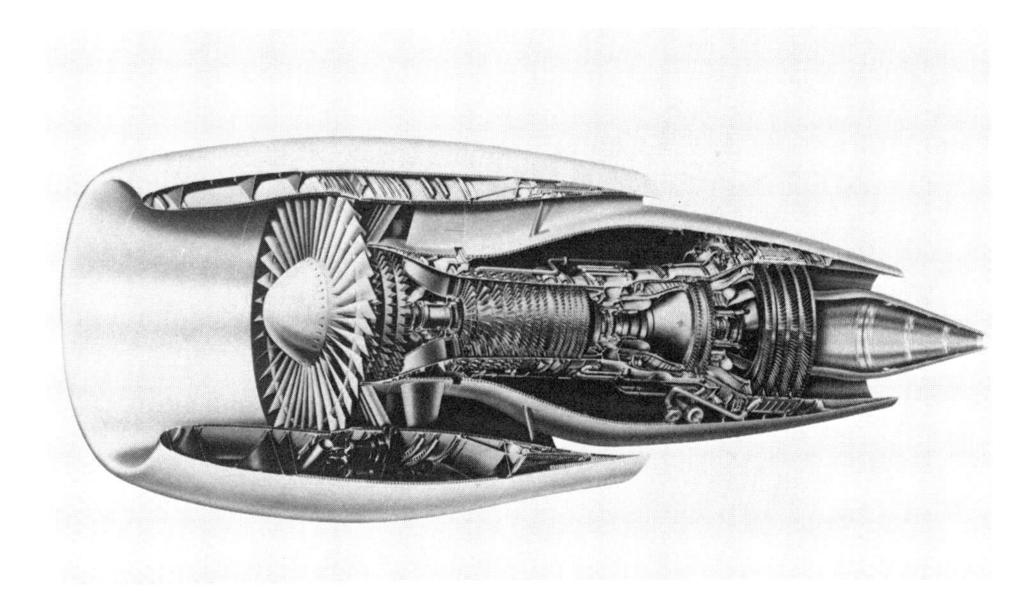

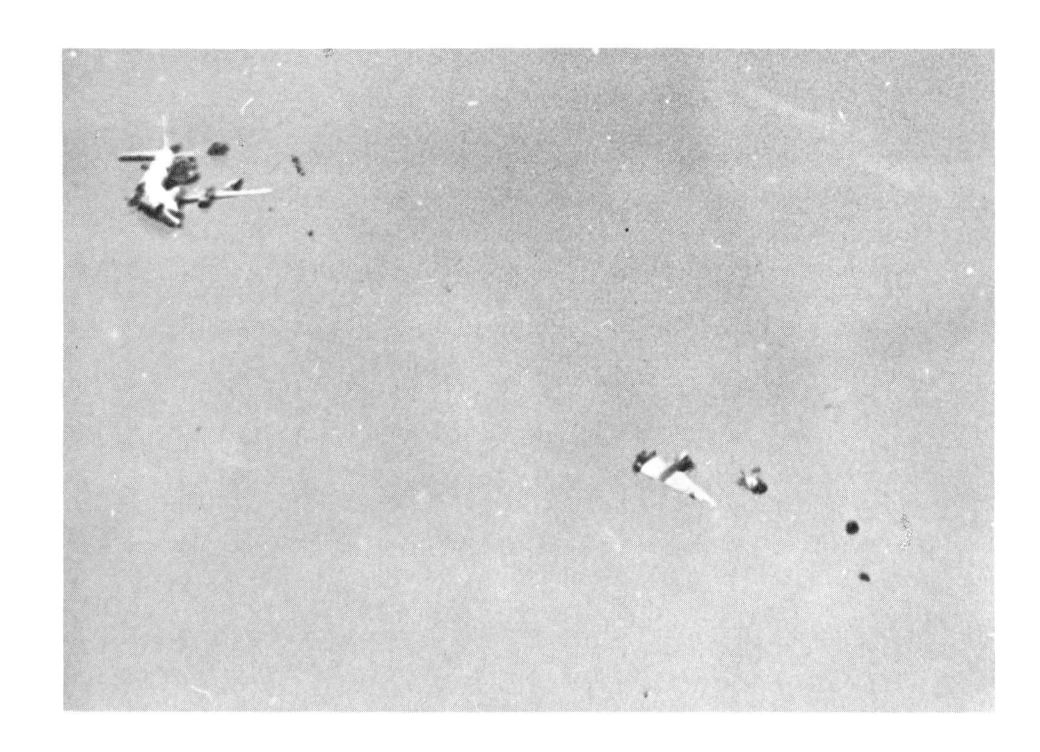

Luckily, Clarice and I were assigned seats near the tail of this DC-6B, which crashed, due to pilot error, into a "dry lake" six miles from Mexico City's airport.

Clarice and I decided to make a run for the Allenby Bridge across the historic Jordan River separating Transjordan from Palestine. The border was twenty miles away, 3,620 feet below Amman, 627 feet below sea level! Without a British visa, we hoped to be able to talk our way across the Allenby Bridge. We started out from the hotel heading ostensibly back toward Mafraq (to fool the hostile crowd)—then swung *Rosinante* around 180 degrees, drove through Amman at 40 miles per hour and headed for the Palestine border on a superb highway. What a relief to be out of that town! It was a clear, sunny day. The view from Amman, overlooking the Jordan Valley and the Dead Sea, the Judean Hills beyond, was beautiful. The curving road passed a sign ENTERING BELOW SEA LEVEL, near a plaque that indicated where Moses hit the rock—and water came out! Another mile, and we were near the Jordan River, a narrow, nondescript bit of water flowing between willows. When we finally saw the Allenby Bridge, we thought we had made it. But something else ruined our dream scenario: There were no British guards whom we could talk into letting us cross the frontier. Instead, Arabs had put up two roadblocks between them and the British border police 300 yards farther down the road, so that there was nothing we could do short of running down the Jordanian guards and bowling over the barbed wire roadblocks. Had we succeeded, it would have been okay. Had we not, we would have been dead ducks. It was another difficult and painful decision for us to make, but it was the only realistic alternative: Return to unfriendly Amman, quickly find the British Royal Air Force base there and ask for protection.

Rosinante performed flawlessly in her climb back to Amman. We raced, for the second time, through downtown, ran a couple of red traffic lights—with crowds after us shouting anti-American slogans. We found the Royal Air Force compound, where the guard raised the gatepost as soon as he spotted the RAF insignia on our Jeep's hood. He then quickly lowered the gate, preventing the crowds from following us. Air Vice-Marshal Dawson, the RAF commander, invited us to stay at his home on the air base until we knew what was what. We deliberated with him and his charming wife how best to get out of Amman into Palestine. Chipsy, in the meantime, had located the Royal Air Force kitchen and made it his headquarters. Vice-Marshal Dawson's suggestion was to abandon

Rosinante in Amman and fly to Jerusalem in an RAF plane he would supply; but we balked: We wouldn't desert our loyal Jeep "just like that." We then agreed that the air vice-marshal would fly to Palestine's capital to get us British transit visas; he would provide us with an armed escort across the Allenby Bridge to the American consulate in Jerusalem.

Four days later, an armored car in front, our Jeep driven by an RAF sergeant in the middle, and a troop carrier with British soldiers and us bringing up the rear left the Royal Air Force compound in Amman for Palestine. Roadblocks were promptly removed by the Jordanians at the Allenby Bridge; the border guards on both sides saluted and let our little convoy pass without inspection. Arab roadblocks on their eastern half of Palestine were not manned because it was Saturday. (Jews were not expected to drive on a Sabbath.) A few hours later we were in downtown Jerusalem, in front of the American consulate in the international sector.

By that time U.S. Consul Roberts had heard about us from the British administration in Jerusalem. He awaited us at the consulate's open front door: a most cordial welcome by a friendly countryman. (He was killed by an Arab sniper, shooting him across the cemetery in front of the consulate a few weeks later, we were told.) Roberts warned us *not* to try the downhill run from 3,000-foot-high Jerusalem to sea-level Tel Aviv because one narrow sector of the sixty-five-mile-long, excellent highway near Ramallah was bordered by steep cliffs at both sides and was a hotbed of Arab snipers. The British were unwilling to dig them out of their positions, from where they tried to control the only access road to Jerusalem from the Mediterranean Sea. Disregarding the consul's advice, we drove off again a few days later, ready to ram through any Arab roadblock outside of Ramallah, halfway between Jerusalem and Tel Aviv. We never got that far: ten miles out of town we were shot at by a sniper. A .30-caliber bullet hit *Rosinante* with a bang, inches behind the last jerry can, finding a home in the travel typewriter case. Whether the gunman was Jew, Arab or Britisher we'll never know. I swung the Jeep around sharply, and we headed back to Jerusalem as fast as *Rosinante* could climb the long hill to town. When we had driven safely all across Asia, it would have been stupid to get killed at the very end of this fantastic trip. We had had enough, all three of us agreed. Back in Jerusalem, we advertised the Jeep in the *Jerusalem Post* and sold her two days later to the American Potash Company, for exactly as much as we had paid for her and the tires in Hong Kong, 10,000 miles ago.

We reserved two seats on TWA's next flight to Paris, due from Bombay in four days, then boarded a convoy of three armor-plated buses—their seats removed to make space for six spare wheels with tires—for its daily run through the blockade, and arrived at Tel Aviv, but not without one more incident. Yes, the convoy *was* attacked at Ramallah, exactly where American consul Roberts had warned us it would be: halfway between the coast and Jerusalem. A brief battle developed with sawed-off Bren guns and hand grenades which young Jewish girls had hidden under their skirts and in their blouses from British military police. Jewish men fired out of the slotted and hinged rear "windows" of the armored box in which we were riding. (Girls hid the weapons because *they* were not searched by British patrols.) Arabs flattened the six tires of each bus with bullet rounds from their submachine guns. One of our passengers was hit in the side of his neck, fortunately not fatally, before we passed Ramallah. The "bus" drivers did not stop for a moment and continued to drive on all flat tires until we were again in friendly territory. Aides mounted the six spare wheels carried inside each bus. We reached Hotel Kate Dan in Tel Aviv five hours after we had left Jerusalem.

Our plane tickets were for one of the last weekly flights of TWA's Constellations to Paris via Athens and Rome. Last, because TWA had announced that it would discontinue flying into Tel Aviv in view of the unrest anticipated when Britain would pull out from its Mandate to govern Palestine three months later. To get Chipsy aboard the plane from Tel Aviv to Paris, we made thorough preparations. We did not mention our Airedale to TWA when we booked the tickets in Jerusalem. There was no question in our minds but that the TWA people in Tel Aviv would refuse to let a good-sized dog fly with us in the passenger cabin. (The plane's baggage hold was not pressurized.) For two nights prior to the plane's arrival from Bombay and its departure for Athens, we made Mr. Chips sleep in our hotel room in a large brown duffel bag with his head sticking out. Clarice had taken from the downtown TWA office a handful of TWA tags normally tied to one's coat and carry-on baggage. She stitched several of these cardboard tags to the duffel bag. We threw away all our worn-out clothing; we thus had nothing but hand baggage . . . and Mr. Chips!

When we arrived at Tel Aviv's airport, Lydda, an hour before departure with no luggage but the dog in his duffel bag under my arm, his head sticking out, the Arab gate attendant stopped us. "You can't take that dog with you in the cabin!"

"Heck," I replied, "we went all around the world with him in this TWA dog flight-bag. We just arrived with him from China; look at our passports! He weighs forty pounds, twenty pounds less than the permitted baggage weight limit for two tourist passengers." The TWA attendant examined our passports and saw the many visas. It never entered his mind that we could have passed through all these countries without traveling aboard an aircraft! He let us go through, saying that the "TWA dog flight-bag" regulation must be a new one of which he hadn't heard yet. He apologized and wrote on the ticket, under baggage weight: "(1) 40 lb. Airedale dog." Once we were in the air, fellow passengers played with well-behaved Chipsy, whom we let out of his flight bag after takeoff.

During taxiing to the arrival gate at the Rome airport, Clarice asked if I'd ever seen that great city. I had not. So, why didn't we stay there and take the flight to Paris next week? "Because of the dog. There will be an American TWA manager at the Rome airport. He will be smarter than the fellow was in Tel Aviv," I replied.

We tried it anyhow. We walked down the plane ramp, with Mr. Chips in his flight bag under my arm, got a fine room in the first-class Hotel Roma, and gave Chipsy a bath in a black marble tub. For a marvelous week we walked around the beautiful city, with the dog on his leash. A Swiss Guard held Mr. Chips while we were sightseeing inside the Vatican. When we emerged, Chipsy was just finishing the guard's sandwich . . . The three of us marched around the Colosseum and King Umberto's giant monument. Wherever we went, Chipsy went with us; the Italians loved him.

The critical day of departure from Rome came, and with it the anticipated problem with TWA's American airport manager: "You can't . . ." et cetera, et cetera.

"Why not? We just came from Tel Aviv, after a trip all around the world with that dog . . . look at the ticket!"

The American TWA official looked. That's exactly what it said: "baggage weight: (1) 40 lb. Airedale dog." "Forgive me," apologized the American, "I hadn't heard about this dog flight-bag before." And into the dark-blue sky went the two Neumanns and Mr. Chips, from Rome to Paris by TWA.

The dog stayed in our room in Paris, sat underneath tables in sidewalk restaurants, visited the Place d'Étoile and the top of the Eiffel Tower. Chipsy celebrated his second birthday at the American PamPam restaurant, where the French chef, in spotless white uniform and high hat, served him a bowl of thick soup on the floor in the middle of the restaurant, before a group of applauding customers.

We didn't have enough money to fly from Paris to New York. Therefore we booked on the much less expensive French liner *De Grasse* from Le Havre, the French port on the Atlantic Ocean, to New York. Although we had left ourselves two full hours to get from our hotel to the boat train station in downtown Paris, we nearly missed the special train to Le Havre. There were only a few taxis in Paris in early 1948, and each of them seemed to have fares. To attract any vehicle, we stood on the sidewalk frantically waving at passing vehicles *all* the foreign currency notes we had left over from the trip. A beer truck loaded to the hilt with wooden barrels stopped. Clarice quickly climbed in front between the driver and his assistant, while I lifted Chipsy on top of the barrels and hurriedly climbed after him, ducking deeply as the truck raced underneath the low-hanging traffic lights in Paris. We made the boat train literally by seconds—running as fast as we could from the station ticket window to the train platform.

Once aboard the *De Grasse,* Mr. Chips stayed in a kennel next to the ship's kitchen during the twelve-day sea voyage, except for two daily walks with us around the deck. It was a wild and rough trip across the stormy March Atlantic, the liner wallowing in the huge waves. And once again, as when I returned on the troopship from India two and a half years before, I was seasick. . . .

The New York immigration officers wanted to hold Mr. Chips in the usual six-month quarantine for newly arriving dogs. But we had a rabies certificate issued in Shanghai and printed in Chinese except for the dog's and animal hospital's names, and interesting-looking seals and Chinese stamps. We convinced the official that this document—which he could not read, of course—permitted the dog to enter the USA without quarantine.

We were back in the States. A taxi took us to Grand Central Station, but had to pull up at a curb halfway there: Mr. Chips was carsick in a New York taxi!

15

Just Lucky, I Guess... at General Electric

There was no response from General Electric to my job inquiry mailed three months earlier from Teheran. I was looking forward to returning to Douglas Aircraft in California and hoped that their year-old "You have a standing offer to return" was still valid. But because Clarice preferred to remain on the East Coast, I phoned GE in Bridgeport and was told, "Yes, we have forwarded the letter with the pretty stamps to Lynn in Massachusetts, where our jet engines are built."

It was in March 1948 when the last of the three interviewers at GE bade me good-bye the following day in Lynn. I wasn't quite sure if I had missed something: No question about the salary I expected? No job offer? (The three men could not have failed to recognize that I had a more extensive and practical background with aircraft engines in combat—no jets, though—than had any of their design engineers.) Instead, an assurance was given that I would hear from GE "definitely within thirty days."

"Thanks," I said, "I'll be in California by that time." I expressed my expectation for a definite yes or no, right then and there.

This shook up General Electric's hiring practice, which called for prior approval from one specialist at company headquarters in

Schenectady. After a ten-minute discussion amongst the three interviewers and their general manager, one of them whispered in my ear, "Don't tell anyone: We *are* offering you four hundred and twenty-five dollars a month—much more than any other beginning engineer is getting."

I accepted. I had no illusions about ever being more than an insignificant somebody in General Electric's huge work force of 300,000 people. My expectation was to remain with GE for no longer than two or three years, sufficient time to learn something about jet engines. Nevertheless, I took advantage of my GI Bill of Rights, got a $10,000 loan at 4 percent interest and bought ourselves and Mr. Chips an old, shingled New England house with a large screened porch up front, with blueberry bushes and many trees for Chipsy in the backyard. We began to set up housekeeping in Greenwood, eleven miles north of Boston.

At General Electric's River Works in Lynn, I sat with a large group of young veterans, most of them graduates of engineering college programs sponsored by the military who had not seen action in the war. We were located in a second-floor "bull pen" in a wartime structure, row behind row of twelve wooden desks each, all facing our supervisor . . . and his voluptuous secretary, who evoked bets amongst us young males if what we saw was real. On the floor directly below us were the engine test cells.

My first assignment in the Aircraft Gas Turbine Division was the testing of *axial-flow* jet engines. These were new for GE, which had produced British-type *centrifugal-compressor* jet engines until then. Centrifugal-compressor engines were much simpler, used fewer engine parts and were of lower cost. But they had a substantial disadvantage. By having a larger frontal area, they caused more air resistance and drag, resulting in lower flight speed and higher fuel consumption.

Axial-flow jet engines were not new in their basic design. The Germans had started developing this type of engine as early as 1935, and flew their first jet plane in 1939, four days before the start of the Nazis' war against Poland. The Luftwaffe was obviously far ahead of America; had even begun—with deadly effect—to use jet fighters against our Air Force over Germany in 1945. Many American flyers were indeed fortunate that Germany ran out of jet fuel during the last few months of the war.

I was directed to report every morning at eight sharp to Engineering Manager Sam Puffer, one of GE's jet engine pioneers, on what had gone haywire on development tests during the previous twenty-four hours. It was an ideal assignment, giving me a unique chance to learn quickly what GE's axial-flow jet engines were all about. I had an exceptionally good opportunity to observe what effects one or another component of a jet engine had on its overall performance, measured by reliability, fuel consumption and thrust output. I became aware of the fact that jet engines have a preference for bad weather, rain and snow: The more miserable and cold the climate, the better jet engines like it!

Quality control was (and still is) a key function in GE's jet engine manufacture, from checks of the purity of raw material to measurements of precise dimensions of machined and assembled parts. Much effort went into the development of extremely sensitive instrumentation to detect flaws in forgings or critical fits when an engine was cold or warm. Even such a simple-sounding matter as the tightening of bolts and nuts to prevent them from being overtorqued or understressed, under the many situations of airplane engine operations, required a great amount of research and testing. I became so fascinated in the overall test arena that any thought of leaving GE in a couple of years began to evaporate.

Let me try to make a jet engine expert out of you, in a few greatly oversimplified paragraphs.

Question: How do jet engines lift a heavy plane into the air?
Answer: They don't! Jet engines, unless of a special design like those for helicopters, provide only forward motion to the plane in which they are installed. This movement makes air streak over the gently curved upper surfaces of the plane's wings. When the air flows fast enough, wings—and the plane—will be lifted upward.

Question: How does a jet engine provide the plane's forward motion?
Answer: Quite simply: Cold air is sucked from the atmosphere through a big opening in the front of the engine and is then expelled—heated and at a very high speed—through the engine's rear. The difference in the speed of the air in and out, multiplied by how much air is involved, provides the forward thrust.

To achieve a high discharge speed, incoming air is squeezed within the engine's compressor (a series of "fans" assembled one behind the other) from 15 to 30 times the air's inlet pressure. Aft of the last fan stage, at the point of the highest pressure of the air flowing through the engine, a fine spray of diesel-type fuel is injected, mixed with the compressed air, then ignited by a spark plug. Once this fuel/air mixture is lit, it keeps on burning without the aid of any artificial ignition. Continuously, air is pushed aft by the compressor into the combustion chamber. The fuel/air volume expands dramatically as its combustion temperature reaches as high as 2,500 degrees F. It is then rushed through the engine's turbine (a series of rotating pinwheels), in which the air expands. These turbine wheels are connected by a shaft to the compressor, which is rotated by them. The turbine's purpose is not only to keep the compression-combustion-expansion cycle going (see sketch page 235) but also to provide electricity for the plane, hydraulic pressure to operate the landing gear, a unit to pressurize the cabin air, and the like.

By the time the air leaves the last turbine wheel, it has used up most of its pressure and part of its very high temperature; it expands completely after flowing through the engine exhaust end, the nozzle. How long does all this compression, combustion and expansion take? A fraction of a second, faster than it takes you to read one short sentence.

One of the many headaches the Lynn aircraft gas turbine people had in April 1948 was the hold on engine shipments of newly finished J47 production jets (at that time the standard military jet engine in America) because the SFC (specific fuel consumption) of these held-up engines was 1 to 2 percent higher than had been guaranteed by GE to the military customers. Thirty valuable and badly needed J47's, sealed in airtight containers, were stacked behind the manufacturing building while an investigation was launched to find the cause for the higher fuel consumption.

Near the test cells were half a dozen 20-foot-high iron standpipes of 2 to 8 inches in diameter and a shack. This little area, I was told, was for fuel-flow meter calibration. By pure coincidence I walked into that shack one noon, my brown paper-bag lunch in hand, and watched an elderly technician working there. I asked him to explain to me how he was conducting the fuel-flow meter calibration. Although he didn't know the hows and whys, he assured me that he was following written instructions, orderly and precisely, and that each instrument had to be calibrated monthly. Eating my sandwiches, I stayed in that shack while the worker went to his lunch. I looked over the setup, his written procedures and test results, merely to satisfy my curiosity. By the same coincidence with which I had walked into that man's shack, I happened to notice that the glass tube fuel-flow meters to be calibrated—filled with fuel when their calibration began—were empty after the conclusion of each calibration run, as they should be. However, after a few minutes had gone by, inches of fuel were visible at the bottom of the supposedly empty glass measuring instruments; this was fuel that had slowly trickled down from inside the walls of the now-empty standpipes. The time it took for *all* the fuel to drain was affected by fuel viscosity, which in turn was a function of the temperature of the unprotected standpipes. Sunshine or freezing rain or snow and ice changed the pipes' metal temperature. Therefore, not *all* liquid in the standpipes had flowed through the measuring system during the critical calibration time (as it should have), resulting in an erroneous fuel-flow rate measured on two electric timers, accurate to within one-hundredth of a second. After lunch I brought this matter to the attention of my supervisor. He immediately passed it on to his higher-ups, who talked it over with the military. The conclusion: The fuel-flow instrument calibration was in error. The thirty held-up engines were permitted to be shipped because they were indeed all right. Our division general manager was delighted. This totally accidental solution of a bothersome problem foreshadowed my career: Just lucky, I guess . . . !

As a consequence, I was assigned to work for a senior designer, Bob Miller. His job was to think of a bigger and better jet engine. All his calculations, sketches and drawings for this new engine were neatly filed by Miller in a single three-inch-thick binder; everything was handwritten; calculations were made on a slide rule or by longhand. (Computers or electronic calculators did not exist in 1948. Today the design of a new jet engine fills more than a hundred volumes of computerized data.) I studied Miller's design book from cover to cover. Here I was given another marvelous opportunity: This assignment got me into the inner guts of a jet engine, from a theoretical point of view. Bob Miller had seven years of design experience, which at the beginning of American jet engine history was the longest time possible. Just as I was ready to make my own design suggestions to Miller, another lucky break happened in my career.

I was told to take over the Building 29G complex in Lynn's River Works, at that time this country's largest compressor research laboratory, a brand-new world in itself. It contained an altitude chamber, refrigeration turbines, control stations, a 35,000-horsepower steam turbine removed from a WW II Navy destroyer and much more. I was to see to it that 29G construction was rushed to completion, to instrument the laboratory for operational and research use and to establish operating procedures. A dozen young engineers were assigned to me, together with thirty-two union workers. The lab was to test research compressors (the compressor is the heart of any engine) at simulated altitudes from sea level to 60,000 feet, from standstill to supersonic speed. This giant multimillion-dollar research facility was being built along one bank of the Saugus River, which sent fresh water into the nearby Atlantic Ocean at low tide but which partially filled with salt water flowing upstream at high tide. Every minute, four thousand gallons of river/ocean water were pumped into the laboratory machinery to cool the steam turbine and the hundreds of gallons of fresh water that were needed—in a separate system—to cool the circulating air heated by the research compressor under test. These thousands of gallons were then discharged as a huge cloud of thick white steam visible for miles.

Building 29G became my world for fourteen months—an exciting, fascinating and challenging assignment. My junior engineers, the unionized workers and I were left totally alone. We developed into a high-spirited bunch, precision-coordinated like the crew of a submarine. We met our schedules, and we learned much: amongst other things, that the operation of our laboratory was affected by the phases of the moon, by spring and winter. The 9 to 11 feet of ebb and flood in the Massachusetts Bay and the Saugus River interfered with the rigidity of the building complex, which rested on cement columns rammed deep into the ground. Amazingly, 29G is still standing and operating well!

Tiny water leaks in our giant hot air/freshwater heat exchangers were causing "snowflakes" to form in the chemically dried air that circulated in the research compressor system and plugged up tiny holes in the pressure probes. These many heat exchanger units, once assembled and brazed together inside a large steel tank, were inaccessible for repair. Luckily, we were able to seal the small leaks in the coolers by using my patent medicine, about which I learned when I was an apprentice in Germany in 1933: horse manure, dumped into the heat exchangers' water.

We were haunted by another operational problem: It was essential that each of fifteen operators of this mass of noisy machinery installed in the crowded basement of 29G observe rows of individual instrument dials located 15 to 20 feet from their related control valves, and make necessary adjustments instantaneously. Instead, it took critical seconds for each operator to figure out first which dial belonged to what valve. To fix this what-goes-where problem, I sent our secretary, Helen, to Woolworth's in town to buy hundreds of feet of two-inch-wide silk ribbons of every color they had in stock. We strung a different-colored ribbon between each gauge and its respective control valve. The laboratory basement, admittedly, acquired a rather nontechnical Maypole festival appearance, but unless the operator was color-blind, the ribbons did the trick! To operate and coordinate this throbbing and screaming machinery —giant pumps, motors, turbines, air dryers, air ejectors—was actually more complicated than to do the testing of the research compressor itself. Just when we reached near-perfection in operating efficiency and every one of us knew what he had to observe, adjust and record during a research test, another one of those things happened which definitely fell into the category of "Just lucky, I guess.". . .

I had to live with GE's policy of rotating each young engineer four times to other plant locations, once every three months. However, I had a special case. Fred Brown, married with one child, a former Marine fighter pilot who had seen action in the South Pacific, was one of GE's new, yet thoroughly mature engineers assigned to me for his first three-month test period. An aide to our division general manager, Marion Kellogg (who helped me a lot in later years and became General Electric's first woman vice president), was in charge of rotating these young engineers from Lynn to other GE plants in the country. When she requested me to release Brown for his second three-month assignment, both Brown and I objected. "Brown stays here! He is not a young kid out of college. He likes it here and he's doing a good job." For someone at my low organizational level to dare obstruct GE's company-wide, time-honored training system of newly hired engineers was unheard of. Miss Kellogg was angry; she complained about me to Division General Manager LaPierre. Within half an hour I was standing in his office, called in to catch hell. The big boss of the engine business blasted me; then C. W. LaPierre invited me to join him for lunch, to talk about GE jets and a specific idea—"variable stators"—I had had as a consequence of the compressor tests I observed. Without doubt this was the most important lunch I ever had in my life. It caused my GE career to become what it turned out to be. Two weeks later, Division Manager LaPierre invited me to another luncheon and presented me with my first recognition by GE: a managerial award and a cash bonus for my work at 29G.

The net result: Brown remained—and did not have to participate in further rotating test assignments. As a matter of fact, he was given my job. I was relieved from the now-operational 29G and told to organize a formal Preliminary-Design function—the first real long-range one the Aircraft Gas Turbine Division had—looking toward future generations of GE jet engines, ten to thirty years hence. A dozen inexperienced but bright-eyed, bushy-tailed engineers were assigned to me. We laid out on paper a new type of jet engine about which I had talked with LaPierre, with a modified type of compressor that had "variable stators."

The function of row after row of stator vanes in a jet engine is to guide the air sucked into the engine compressor and to direct it aft into the combustion chamber. In the then-current type of compressors, all stators were installed in a "fixed" position inside the compressor casing. Such fixed position made it impossible to produce optimum compressor efficiency over the wide operating range and different rotational speeds of an engine. Our aerodynamicists were convinced that engine performance and fuel consumption could be improved dramatically if the angle of the vanes somehow could be varied ten to thirty degrees around their bases during engine operation. To accomplish this, I proposed a system that would let the stators change angles during flight without affecting their reliability. Thus, the name "variable stators" was created.

LaPierre provided the funds for me to design and have manufactured a full-size research compressor with these variable stators, to demonstrate in Laboratory 29G its feasibility, which many other engineers had pooh-poohed. Six months later it was on test. It worked—not only well, but absolutely superbly! GE filed a patent application (owned by GE, but with my name as inventor), which was granted a few years later. Today nearly every jet engine made anywhere in the world, including those of the Russians, incorporates this variable stator feature.

Building the new research compressor did not go all that smoothly, of course; my designers and I made plenty of mistakes, had different technical opinions, encountered jealousy between newcomers and old-timers, plenty of manufacturing delays and cost overruns. These caused me to have appropriate discussions over the phone with a manufacturing foreman whom I told, in no uncertain terms, what I considered him suitable for instead of running a precision machine shop at GE. Following these pretty heated phone monologues this man complained bitterly to my supervisor about the uncouth language of "that damn foreigner."

General Electric had instituted a constructive appraisal system in which each nonunion person's performance had to be reviewed annually by his supervisor and commented on by three of his peers. When my turn came to be evaluated, I was pleased about my boss's generally favorable comments, but shocked by his severe criticism of my choice of words when talking over the phone (his input came from that foreman). True, my language still contained a good amount of Army vocabulary. My supervisor, Frank Warner, urged me to "improve fast—or you won't amount to anything!" It so happened that a couple of weeks after the discussion of my performance I was asked to fly to California to look over a different type of jet engine designed by a group of engineers, and to make a recommendation if GE should try to buy it. TWA's Lockheed Constellation from Boston to Los Angeles stopped at Chicago's Midway Airport. The aircraft I was on had developed an engine problem en route which, the repair crew estimated, would be fixed well within six hours. As was customary in those good old days, airlines provided a bus for delayed passengers to be entertained during the repair period, in this case a visit to downtown Chicago with a stop on the way at the Museum of Science and Industry. I got off there.

Right inside the museum's entrance was an exhibit where one could listen to one's own voice. Mounted on the three sides of a U-shaped table were phones where people stood holding receivers to their ears. "Speak into your phone for 30 seconds and listen how your voice sounds to others as it is being played back," a sign explained the purpose of this exhibit. Perfect! The person at the head of the U was just leaving. I took his place; for a moment I tried to recall my monologue with the manufacturing foreman who had complained about my choice of words. Then, for thirty seconds, I rattled off the juiciest expressions that I had picked up in the Army and listened to the recording mechanism playing them back. Not only did my voice sound much squeakier than I had thought; I also hated to admit that it sounded much more profane than I had intended. One foul sentence followed the other. I really sounded awful. Suddenly, I noticed various people on both sides of the table looking at me in horror as they held their phone receivers a few inches away from their ears, having listened to some expressions they had probably never heard before. Through some crossed wiring, I assume, the other phones were connected to mine. I fled aghast from the scene.

In May 1950, GE's Management Association invited me to give an after-dinner talk to two hundred engineers about my war experiences and the three Neumanns' Jeep trip across Asia. I had a captivated audience for ninety minutes. Suggestions were made at the end of the evening that I ought to get myself listed in a professional speaker's bureau. I did, and got more offers than I could handle after working hours! Soon butterflies in my stomach before giving a talk disappeared; I learned to speak without notes before any size group. The modest speaking fees I received certainly came in handy, but the main benefit I derived was the self-confidence to make presentations in exactly as many minutes as the program chairman had called for.

The same year, international postwar relaxation was, once again, nothing but a memory. Tension had begun to build up between the Western Allies and the Soviet Union over the status of Berlin. The American aircraft industry went into high gear, producing a large number of jet fighters and six-engine jet bombers, most of them powered by General Electric's J47 jet engines. Our Lynn plant was too small for the large engine orders on the books—with no room for factory expansion. Even large subcontract orders to the automobile makers Studebaker and Packard were unable to produce enough parts to fill the sudden demand. GE bought a large part of the vacant former Curtiss-Wright aircraft engine war-surplus plant at Evendale, near Cincinnati, where tens of thousands of propeller engines had been produced during World War II. The empty plant was one and a half miles long and contained the second-largest manufacturing building in the USA, so large that we peons could not imagine what we would ever do in such a huge facility. The story circulated that we would stock it with mountain goats—for a mysterious hunter, Mr. Jack Parker! Most of us had heard about a Jack Parker working with GE on nuclear matters, a man who was responsible for the assembly of Liberty ships at Todd's shipyard at Houston during the war, in record time.

Most of the Aircraft Gas Turbine Division was going to be transferred from Lynn to Ohio. The Neumanns, like many of their friends, hated to leave the beautiful East Coast, but there was little choice. As an attraction for me to move, I was offered the designer-in-charge job for the world's first nuclear-powered aircraft jet engine (not the reactor itself, which was done by a specialist team). The ANP (Aircraft Nuclear Propulsion) Department was under General Manager Roy Shoults, who had been deeply involved in America's very first jet aircraft, the Bell XP-59, which flew in the California desert in 1942. Shoults also was the engineer sent to England by the Chief of our Air Force one year earlier, in 1941, to conduct negotiations for GE to build a British jet engine under license. During 1951 while I worked in his department, he gave me much good, practical business advice. The best: "Walk the shop whenever you can!" He insisted—and I know he was right—that a shopwalker will return to his office wiser and better informed than if he has *not* taken the time to look around and talk to people other than his staff.

Why did America want nuclear aircraft propulsion? Because the range of a bomber powered by such engines would be virtually unlimited; only the crews' well-being would determine whether the plane would circle the globe once, twice, or many more times. When I talked to a crew that had lived for many days in a simulated bomber cockpit and crew quarters, I learned that their main complaint was "sticky fingers after peeling oranges"! The disadvantage of a nuclear-powered plane would be, of course, the potential for a major catastrophe if there ever were an accident.

The engine was designed to operate with either conventional or nuclear fuel. In 1954 it ran on both fuels in a test facility in Idaho and was considered fully successful. My association with the nuclear program gave me the chance to visit the atomic facilities at Oak Ridge in Tennessee several times, and to learn something about nuclear energy. At the same time, much lower fuel consumption of future conventional jets—thus increasing a plane's range—was visible on the horizon, and promised to continue to get still lower during the next ten to fifteen years. That fact, combined with a newly developed aerial refueling technique from flying tankers, eliminated the urgency for a nuclear aircraft power plant.

16

The Amazing GOL-1590

On a Saturday afternoon in September 1952, just as Clarice and I finished playing badminton on the scrawny lawn of our rented home in Cincinnati, Division Manager LaPierre phoned. "How about coming over to my office, just the way you are?" Roy Shoults, the manager of the aircraft nuclear engine program, was sitting there. At my low organizational level in 1952, LaPierre was a demigod. "Gerhard," said he, "I want you to start Monday morning with the design of a brand-new Mach-two [twice the speed of sound] jet engine using your variable stators. Pick a team of your favorite engineers and manufacturing people. In thirty days we are going to hold a competition between your design and one laid out by another team."

The winning engine of the two fundamentally different designs would determine the characteristics of the next generations of GE engines. Much of the design layout of any new engine, even today, is a matter of intuition or "feel" by its chief designer—just as it is with a piece of art or music. (I don't know anything about computers, unfortunately, but I doubt very much if *any* computer system can develop, by itself, engine simplicity or the optimum arrangement of maintenance features, for example.) LaPierre wanted GE's new family of engines to be superior in performance and reliability, lighter in weight, smaller in diameter, simpler in manufacture and easier to maintain in service than any competitor's product, he explained to me. I was excited about the tremendous opportunity just given me, less than five years after I had joined General Electric as a test engineer.

The paper competition took place in a hotel at French Lick, Indiana. At the end of two days of presentations and detailed cross-examinations, arguments flying pro and con, my team's paper engine with its variable stators was declared the winner! As manager of the VSXE (variable stator experimental engine—promptly called by my wife "the Very Sexy") project, I was given only one year to design, build and bring to test a full-size demonstrator engine which was to prove that what I predicted *could* be done was indeed doable. Normally, such a job takes from eighteen to twenty-four months. My team of 120 engineers, draftsmen, secretaries and material and manufacturing specialists worked in one common bull pen six days a week; I moved out of my adjacent private office to sit with the troops. We held design reviews several times a week, with key engineers participating. We were also lent space in one of Evendale's large manufacturing buildings for the assembly of hardware. I insisted that this area be painted snow white, in stark contrast to the customary unattractive dark factory gray (this was long before the medical cleanliness of the space age became a matter of second nature). Ceiling lights were lowered and tripled; tarpaulins from ceiling to factory floor sealed off our bright area from the rest of the plant. Anyone working in this "white area" could not help but get a feeling of care and importance.

During the assembly of the twelve thousand individual parts making up one jet engine, all of us were one team: workers, managers, foremen and engineers. The designers had as many black fingernails and bruised knuckles as the union workers next to them. When I had to attend a black-tie reception one evening, I sneaked away after the dinner to be with the night shift in the GOL-1590 (as the VSXE was designated by the government) assembly area—in my monkey suit. Nobody thought this strange. It was natural for me to be in the work area at any time, day or night. Our morale was superb. Each morning I stood on top of my desk and got the attention of my associates by ringing a Swiss cowbell presented to me by Jim Krebs, one of my brightest engineers. I updated the assembled team on what we had accomplished during the past twenty-four hours, what unforeseen problems had arisen, and how many days we had left before the next milestone was due in our tight schedule. During the year's tireless effort we had no sicknesses, no absences from work by any of the 120 people working on the GOL-1590 demonstrator engine. Vacation time was voluntarily transferred to the following year by everyone.

It was the best-carried-out engine development program I have ever seen, with as small a team as possible. Each member had been handpicked, and each was overloaded with work and could not afford to diddle around with decisions that had to be made on the spot. Drawings were hand-carried into the machine shop; orders for hardware were placed over the phone rather than by mail. The whole project operation reminded me of the Flying Tigers in China, who were also undermanned, overworked . . . and successful! I thought many times how proud General Chennault would have been to see our GOL-1590 team in action.

A non–design-related matter typified the *esprit de corps* of our design group: Cal Conliffe, one of our engineers, was black; a good, hardworking, reliable man well liked by everyone. We also had a stunning-looking white secretary with the cutest southern drawl. When our team decided to take a few hours off for a brief pre-Christmas party at any hotel or restaurant in Cincinnati, each dining place refused to accommodate us when we told them that we would have one black engineer with us. Conliffe offered not to participate so that the rest of us could have a party. It was our southern gal who climbed onto her desk and proposed to all that "none of us goes if Cal doesn't!" Everyone cheered. This seems hardly worth noting today, but thirty years ago this was indeed a landmark. (We *did* find, ultimately, a veterans' hall that opened its doors to *all* of us.)

One week before Christmas 1953, shortly before midnight, engineers and testers pushed the first GOL-1590 high-performance jet engine, its air inlet painted with the Flying Tiger shark's mouth as a Christmas surprise for me, into a specially prepared test cell. It was by far the most powerful jet engine GE had ever built, not only in absolute thrust but also in relation to its small size and low weight. A cardboard attached to the engine read: FIRST DAY—FULL SPEED OR BUST! At four o'clock the following morning (with twenty hours of work already behind us) we hit the starter button.

The engine fired instantaneously and gained speed so unexpectedly fast that, unprepared for such an unusually short starting cycle and such rapid acceleration, we lost control of the variable stator operation, which was done manually during the first few tests. The engine shook violently. A steel bracket ("dog bone") supporting the 3,500-pound engine hanging from the overhead test stand broke and let the front end of the engine drop. It bounced off the heavy test frame; smoldering pieces of the VSXE lay all over the test cell floor. We were too tired to cry. . . . Yet we had met our two commitments: The new GOL-1590 did start instantaneously, accelerated very fast, and busted—all during the first day!

Terribly disappointed, we cleaned up the mess after giving ourselves a few hours' sleep, gathering our wits. We began to rebuild the engine and to replace the defective mounting. Most of us worked over the Christmas and New Year holidays. Three weeks later we had the engine back in the test cell. This time we were prepared for an instantaneous start. The demonstrator engine ran beautifully, exceeded full power in eight hours, met every commitment we had made—and bettered them! Within two months Engine #2 went to test as #1 was pulled out of the cell to be thoroughly inspected. Engine #2 operated under various atmospheric conditions. Then came Engine #3. Sure, we had our share of miseries and headaches, misfits, bum suppliers of raw materials and forgings, seals, rubber hoses, bearings, controls or whatever else we bought from the outside. Even components made in our own shops were not always trouble-free. But we overcame. . . .

Four months after work on the GOL-1590 had begun, a government contract was offered GE to develop the GOL-1590 for production in a somewhat smaller size, under the Air Force designation J79. The latest Air Force and Navy fighters, supersonic bombers and missiles were to use the J79. GE was to test-fly the engine initially and to produce the first batch of several hundred engines. Five altitude and speed world records were broken with the J79 within a few months. Neil Burgess, a brilliant engineer and experienced project manager of the older, successful J47, had been put in charge of the J79 project. Between 1955 and 1982 we—and our licensees in many countries around the world—produced over 19,000 J79's worth more than four billion dollars. This GE engine is still being built today (1984) and has the documented reputation of being the most reliable military combat jet engine ever built. It is used not only by our Air Force, Navy and Marines, but by all of NATO, the Israelis, Japanese, Koreans, Taiwanese and many more. It has seen combat in many parts of the globe. Yet the J79 engine, which used the variable stators for the first time, was only the tip of the iceberg. Billions and billions of dollars' worth of GE engines with variable stators were, are and will be coming off production lines to power airplanes around the globe, big ones and little ones, military and commercial.

In 1953, General Electric, which claimed to make more different products than any other company in the world, underwent a major reorganization. Its president, Ralph Cordiner, decided to decentralize the company and to divide it into fifty departments which were self-contained businesses, each headed by a general manager. Accordingly, the Aircraft Gas Turbine Division was divided into five departments: I became a section manager in one of them, the Flight Propulsion Laboratory Department, and reported to a very good general manager of the "advanced thinking" type, David Cochran. I was responsible for the Preliminary Design Engine Section, with the charge of thinking of our next steps to be taken—to assure GE becoming the unchallenged front-runner in engine technology.

GE's outstanding successes with the GOL-1590, the J79, the nuclear aircraft engine and many more of our advanced programs would likely not have happened without the help of two extraordinary people.

One was Morrough P. O'Brien, the former dean of engineering of the University of California at Berkeley, a real engineer. His responsibility as a consultant was to earn the confidence of our designers at all levels, visiting them individually at their desks or having lunch with them, in order to ferret out (not solve) problems which otherwise might not have surfaced. (O'Brien continued to consult for me for twenty-five years.) Although he never admitted it, I credit him with having urged LaPierre to transfer me from the less urgent nuclear engine program back into the mainstream of the jet engine business. The internal engine competition of 1952 set the course for the technical direction in which General Electric's engines were going to head—and also for my own career.

The other prominent person was a genius: Herb Grosch, who set up a massive computer operation in a separate building in Evendale in 1954. He foresaw the incredible power of computers before almost anyone else at General Electric or anywhere else in industry recognized it. Grosch trained many young women in the science of computers. His computer building in Evendale was called by those of us who did not understand what was going on inside "The House of Ill Compute."

In 1954, Clarice and I were asked to make a goodwill visit to Savannah, Georgia, on behalf of the General Electric Company. I gave a talk before Savannah's Chamber of Commerce on my war experiences and our Jeep trip across Asia. A couple of months later, I received a letter from a Mr. W. L. Bond, postmarked Warrenton, Virginia. Mr. Bond wrote that a friend of his had heard an exciting tale about the Orient at a recent Chamber of Commerce meeting in Savannah. Could it be that I, the speaker, was the same young man whom he had happened to meet in Hong Kong in June of 1940, escorted by a British soldier, and whom he had helped to slip out of British internment there? The letter writer turned out to be the same Mr. Langhorne Bond, vice president of Pan American Airways. Clarice and I visited Mr. & Mrs. Bond on their farm near Washington, D.C., a few weeks later—and several times since then. It was and it still is a thrill to meet again this old China hand who had helped me into freedom and—indirectly—to America.

Most members of the GOL-1590 team remained with GE's jet engine business. In 1973, twenty years after the initial test of the demonstrator engine whose program had broken all kinds of records, we held a reunion in Cincinnati, for which the clever wife of Jim Krebs (now a corporate vice president in the Aircraft Engine Group) composed this witty poem:

Owed to the 1590

Greetings, friends, let's celebrate
The 1590's birthday date.
We'll honor twenty years tonight
Of GE's high-thrust, low-weight might.
To the 1590!—The way to go
Two times supersonic flow!

But in our merriment, let's pause
To wallow in some self-applause . . .
To toast that grand old Turbojet,
And in the process, credit get.

All hail the 1590, then
Designed by most ingenious men.
Conceived to beat Pratt's dual spool,
Beat their weight, consume less fuel.
The edict was, "Go full speed, boys,
Hang the smoke, and damn the noise!
With single-spooled high pressure rotor,
Create a money-making motor!"
Lift your glasses, drink your bourbon,
Salute the three-stage axial turbine!
All hail you eager innovators
Of high-pressure ratio variable stators,
Thin discs and lever actuators
Who built and ran three demonstrators!

—Mitch Krebs
December 12, 1973

17 Getting to Know GE Jet Engines

I don't know how other companies handle appointments to high positions. Nor can I even say how it is generally done within General Electric. But I know what happened to me, and it certainly was different from what one reads in managerial textbooks. It was March 1955. Clarice and I had just gone to bed when I received a call from the Cincinnati airport: Group Vice President for Atomic Energy and Defense Products LaPierre, and Parker, who had joined our engine business and had become its division general manager, would like to get together with me in forty-five minutes. We met shortly before midnight in a downtown hotel lobby, where they informed me that I would be named general manager of GE's Jet Engine Department next morning. My salary would be doubled. Not only was I totally surprised, but I was totally unprepared for the new job of being responsible for five thousand people. I had never attended the company's management school or any of the fashionable university management training seminars. Overnight I became the manager of many people for whom I had worked for years. This touchy relationship switch was the only concern I had in taking on the much broader responsibilities. I guess that it was my good batting average in doing what I had committed to do that made high-up management choose me as department general manager over so many others who had more years of experience on jet engines and service with General Electric.

A week after I had taken over the new job, I learned from my labor relations manager (whom I had inherited as part of my staff) that a union/GE management meeting discussing strike issues in my department was scheduled for the next morning. Where? "Don't bother about it, Mr. Neumann, it's all set!" What was all set? I was told, "Such matters are handled by us professionals." Of course, his cocky answer caused me to attend that meeting, whether my relations manager liked it or not. I had to force myself to sit quietly in the corner of the meeting room observing the proceedings, and was shocked at the predetermined, inflexible attitude of my representatives. They actually had their minds made up before the meeting how many cases on the agenda the company was going to "lose" and how many it would "win." I could barely wait to get back to my office and start the fire-works! Those did occur, I can assure you! And they included the relations manager (pun intended!).

I quickly came to know my new staff, their professional talents, ethical standards, leadership, judgment and degree of enthusiasm. Those inherited members who "would be better off somewhere else" did not remain with my organization. It did not take much time to build a team of top-notch managers selected by demonstrated performance rather than seniority who were definitely not yes-men but still would carry out decisions forcefully, even if they personally were in disagreement with them. I hired a retired Navy accountant who was highly intelligent and very stubborn; I assigned him the job of questioning me about all major business decisions before I made them final. Schmidt was paid to be a professional S.O.B., and he did a great job being one! We had lots of arguments but personally I was always very fond of him.

With my new position came perquisites and a fancy new office. However, I refused to have a rug in my office or a private john next to it—both seemingly essential for all executives in America's big industry. Instead, I had carpenters build an oval-shaped conference table, designed so that everyone in my staff could see each other without twisting his neck. To break off excessively lengthy debates or keep us from straying off the agenda during conferences, I had electric bell buttons installed below the tabletop, one for each member of my staff including myself. Anyone could discreetly press his button with his knee, causing a buzzer to sound: It was the signal for each of us to stop talking promptly, whatever the subject. I was buzzed just as often as anybody else, and it worked well.

It is certainly not a discovery of mine that the most important ingredient for any correct decision, whatever the subject, is to have the facts straight. In the late eighteenth century, German General von Steuben (who trained the U.S. militia) coined the phrase *"Eine Gefahr die man kennt, ist keine Gefahr."* (A danger which one recognizes is no danger.) He expected, of course, that one would promptly take corrective action to eliminate such recognized danger; he wanted to impress on people that any surprise might well constitute the greatest danger. I, too, made every effort never to be surprised. I tried to find out whatever I could from every level of my own organization or from the customer. Informal visits and discussions with my workers and engineers during first, second and third shifts became routine. My staff and I agreed to keep ourselves and also my boss, Jack Parker, promptly informed of good news or bad so that there were no surprises at upper levels in the company; to run a democratic organization (in which I held 51 percent of the vote); to be fair to our workers and to make every effort to make them a part of our successes as well as failures.

To keep each of us key people *au courant* and have the best chance of making correct decisions based on facts, I established an I.O.I. (items of importance) system: A daily, typed single-page memo covering the main events of the past twenty-four hours was mandatory, from everyone reporting to me. Copies were distributed to the other members of my staff. I myself wrote I.O.I.'s to my boss, Parker, with copies to my associates. (To delegate the writing of an I.O.I. was verboten.) The *written* word in our I.O.I.'s demanded an extra effort from the writer at first, but it soon became second nature and had many advantages: One could read about important happenings while flying, or sitting in a hotel room; each staff member was more thorough in checking his facts before putting them down in black and white than if he merely talked about them over the phone; there were fewer chances of misunderstandings; debates whether or not a certain matter "had been mentioned" did not occur since the summary of any significant verbal communication *had* to be recorded in an I.O.I.

I also thought it useful (and so it turned out to be) to inform each of my employees annually in a series of mass meetings of the "state of our nation," i.e., our Jet Engine Department, followed by questions from the floor—and answers. Once a year we rented a large circus tent, seating two thousand people at a time. My staff and I held several two-hour sessions attended by everyone in my department on each shift. We reported to them—secretaries, management and laborers sitting side by side—what our organization had accomplished during the past twelve months, where we beat our competition, why we were defeated ourselves, what the customers thought of our products and their quality, and what lay ahead of us in the next twelve months. I coined the phrase "Do it right the first time!" To put a little spice into one of our "Father Neumann's revival meetings," as they were called, six of our prettiest secretaries volunteered to parade in short-shorts and tight sweaters—remember, this was in the fifties!—down a runway built a few feet high through the middle of the tent (this was my wife's idea!), each girl carrying one word of that phrase mounted on a cardboard. When the good-lookers showed up on the stage, their message boards were deliberately scrambled; the girls paraded about until the words were placed in their proper sequence. (We found that some of our men attended several of the sessions during that meeting—and I am sure it was *not* the business message that brought them back into the tent.)

In 1956, the North Atlantic Treaty Organization decided to permit the Germans to rebuild their Luftwaffe, scrapped in 1945. When the German minister of defense, Dr. Franz Joseph Strauss, accompanied by a group of World War II top Luftwaffe generals, came to America to select the first postwar fighter plane which Germany intended to purchase in large quantities, it turned out that each of the planes under consideration by the Germans was powered by one or two engines made by GE. On the day of their visit to Cincinnati, our U.S. secretaries of defense and commerce joined the Germans in my conference room in Ohio for a briefing on the engines. Presentation charts had been prepared in German; I personally made the technical and sales pitches in their language. To my great embarrassment, I had difficulty in speaking German fluently, because there existed in 1956 many German technical terms pertaining to jet propulsion that were not around when I lived in the old country. After the German visitors had left, Jack Parker gave me a friendly pat on the back. "We knew that you couldn't speak English, but now we know that you can't speak German either!"

In October 1958, I was transferred back to my old starting location at Lynn, Massachusetts. I was to take over the department Jack Parker had launched in 1953. SAED (Small Aircraft Engine Department) designed, developed and produced small jets for training planes and small fighters, and small gas turbines (similar to jets) for helicopters. The department had slid into a managerial problem after Parker had left for Ohio in 1955, to become our division general manager. Within one month I replaced eight of the ten managers I had inherited in SAED. In addition, 25 percent of the engineers, management and shop labor had to be laid off to revive the failing business, whose costs exceeded contract terms. My personnel relations manager, Bob Miles, gave me smart advice: Since we had to lay off more than a thousand people, "let's do so all at once," he suggested, rather than drag this unpleasant task out piecemeal. Those who had to be discharged were better served if they got the bad news quickly so that they could look sooner for a new job.

To drive home to my people that a fresh wind was blowing, I advanced office starting time from 8 to 7:30 A.M., ending working hours half an hour earlier. The earlier starting time established an atmosphere of urgency. When waiting lines at food automats inside the plant became too long and coffee continued to be spilled on floors and stairways despite plenty of warnings, I had all food and drink vending machines removed, in spite of a personal phone call I had received from GE's chief executive, Ralph Cordiner, who warned me of a possible consequent work stoppage. This did not happen. I began to encourage competition by outside manufacturers with our own shops; henceforth, our own production manager did not automatically get orders to make hardware for our own engines if he was noncompetitive with outside manufacturers.

Lynn was the company's second-oldest plant. Many poor working practices had been firmly embedded there for decades. A showdown between tough labor and toughened-up management was inevitable. It didn't take long to materialize. The first serious work stoppage in my reign occurred in 1960. On the strike's first day, I was frustrated (i.e., pissed off!) about the lackadaisical attitude of our Lynn GE employees, the city's police, newspaper reporters and the local population. All were accustomed to GE's employees *not even attempting* to cross a picket line ("That just isn't done here in Lynn," I was told). Prior to the threatened walkout, I had made it clear that "this plant stays open. Whoever does not show up at work will not get paid." Period! That was about as straightforward as one could put it. When the strike started, Lynn's police did nothing to stop illegal mass picketing at the plant gates until we secured a court injunction against the unions blocking entrances.

The plant indeed remained open; more and more nonunion employees, engineers and secretaries crossed the picket lines by car or on foot, past the hollering and jeering strikers. I held daily meetings in the plant auditorium with those who came to work. A few union workers, too, wanted to enter the plant but were afraid of their associates, so they hid in trunks of cars and were driven through the picket lines. After one week we had an attendance rate of nonunion personnel that was better than it was on the average no-strike days: 96 percent of those ineligible to strike came to work! I received a number of personal phone calls at home from union workers loyal to GE who did not want to strike—and who asked me to meet with them illegally. Even a top union official "happened to be" at the pier where I kept my sailboat, on Saturdays and Sundays. He wanted to discuss the strike matter "off the record." Management had the upper hand from the first day on, and the workers knew it.

When pickets cut tires of automobiles driving into the plant, a group of volunteer engineers organized themselves inside the factory to mount new tires flown in by chartered helicopter or brought across the adjacent Saugus River by motorboat. We promised our nonstriking employees to repaint, at company expense, car fenders and bodies allegedly scratched by pickets, after the end of the strike. (I've never seen so many old and beat-up cars enter the plant gates!) One morning at four, I received a call that tar was being spread across all entrances to the plant and thousands of large tacks were being stuck head down into it. The tacks virtually guaranteed four flat tires for every car driving across the tar. A few of my associates and I entered the GE plant through a back door by 5 A.M., loaded a truck with sandbags and backed it to the picket lines, where we intended to spread the sand on top of the tacks. Pickets were watching our efforts silently from a few feet away. But we had forgotten to take along knives to open the stitched bags. I asked the pickets if anyone had a knife he would lend me. Sure enough, two men came forward: "Here, Mr. Neumann . . ." to the cheering and jeering of their fellow strikers. Two weeks after the nail episode, we distributed to all nonstrikers tie clips and earrings decorated with the nails the strikers had used. In spite of a lot of noise by the pickets, they were good fellows and the majority wanted to go back to work. As I was driving along the picket lines, many workers assured me that "*We* want to work—if we only could. . . ." After several weeks, many of the strikers' wives had enough of their husbands sitting at home; the women organized a march to Union Hall, urging them to end the strike.

During the six weeks before the work stoppage ended, we learned to live without the majority of union workers, to manufacture parts with supervisory personnel, and to assemble engines. In the process of doing things ourselves, we learned how much faster —and with fewer people—certain work processes really could be done. As our test cells began again to belch clouds of white steam (from water-cooled engine exhaust) and it was evident to every picket outside the plant fence that our secretaries had, once again, become Rosie the Riveter of World War II fame, showing up at work in slacks and sweaters and beginning to learn how to machine and weld parts after a rush training by delighted shop supervisors, the strike ended. The "old Lynn plant" attitude was gone forever.

Lockheed Aircraft in Marietta, Georgia, invented a catchy name for their quality improvement program: Zero Defects. My Small Aircraft Engine Department, too, made a major effort to eradicate design and machining errors that cost us 10 to 15 percent each year. We obtained Lockheed's permission to copy their slogan, rented the Boston Garden (a large indoor stadium) for half a day, chartered buses and even subway trains and moved five thousand workers to and from downtown Boston. To promote the Zero Defects program we had speakers from management, unions, General Electric's board of directors, military pilots who flew GE-powered planes and federal officials. The enthusiastic participation by our people and their performance in bettering quality was amazing. We announced tough production and quality goals—like making and shipping one thousand small J85 jet engines in a single year, an impossible figure only a few months earlier; and we proudly celebrated the shipment of the thousandth engine by December 15.

Things were going well. We decided to spread our wings. The business-jet plane market was launched and provided an opportunity for us; we captured 85 percent of it in the sixties. We powered every commercial jet helicopter around the world. Each President of the United States beginning with John Kennedy helicoptered (and still does) with GE's engines. We branched out into nonaviation application, using modified gas turbines derived from our jets. We obtained a contract for standby emergency power for San Francisco's telephone company; pumped Shell oil from under lakes in Venezuela; moved trains of the German *Bundesbahn* (railway); hauled copper out of mines in the Midwest; propelled Grumman's super-fast hydrofoil missile ships. Our Small Engine Department sent service people all over the globe, from northern Canada to Australia. An essential part of our success was, of course, the full backing we received from GE's top management in New York. Especially Jack Parker, our division general manager, understood what we were doing and obtained the needed financial support from GE's chief executive and the directors, who invariably showed a most positive attitude toward our risky jet engine business.

One Friday afternoon, hours before the start of the Easter holiday, we ran into a severe subcontractor problem whose solution called for a miracle; my staff and I sat tensely around our conference table to resolve it. "Isn't there any one of you guys who can pull a rabbit out of a hat?" I asked. Everyone was silent. "If you can't, I can!" Under the table, in my hat resting on my knees, was a live white baby bunny with pink ears which I had bought during lunchtime to take home to my two young children. I slid the hat into the middle of our round conference table and pulled the astonished bunny by his ears out of the hat. . . . We *did* find a solution to our problem that afternoon!

In March 1961, once again without the slightest indication beforehand, I was told by Parker that I would be taking over his job as general manager of the Flight Propulsion Division, effective immediately; he himself had been promoted to the position of group executive of the Aerospace and Defense Group. The turnover of my Small Engine Department to my able successor Ed Woll, one of the two managers in my inherited staff whom I retained after I took over in 1958, required less than a minute.

18

Beating Out the Competition

My new responsibility was GE's *total* jet engine business: large and small engines (military and commercial), aircraft accessories, research and development, marine and industrial; separated into five departments, each with its own general manager. In 1961, my division had twenty thousand employees and was still growing. Although the majority of them worked in the Evendale plant, near Cincinnati, I had no intention of moving out of our unique Swampscott home, four miles north of the Lynn plant, where we had been living since 1958. We had two children, and with dog and cat were beginning to fill up the beautiful large Tudor house overlooking Massachusetts Bay and Boston. A third child was on the way. General Electric's top brass approved my moving the division's headquarters and half my staff from Ohio to Massachusetts; all other Evendale employees remained in Ohio. A major change in my modus operandi was now inevitable: The much more challenging job, with its far greater responsibilities and worldwide coverage, meant much more travel. The then-current trend for decentralization within General Electric had caused expensive duplication of production facilities in our jet engine business. Even more significantly, my five departments had the same prime customer: Uncle Sam. Internal competition is healthy when limited and well controlled, but our division seemed (and indeed was) frequently incoherent and uncoordinated when some of the departments made their independent proposals to our government containing inconsistent approaches.

I suggested to Jack Parker that we recentralize our division. My proposal, counter to GE's publicly declared philosophy at that period, was based on one simple argument: Our Flight Propulsion Division produced only one type of product (in different sizes and for different purposes). Each jet engine had fundamentally the same components: compressor, combustor, turbine. I forecast substantial savings if we could specialize our shops to produce certain types of engine components, albeit of different sizes, for all engines—whether they were assembled in Massachusetts or in Ohio. We could also standardize the engineering technology used in our engines, and thereby utilize limited research and development funds more effectively. Surprisingly enthusiastic, Parker and GE's chief executive, Ralph Cordiner, concurred. They even placed one of the four corporate twin-jet executive aircraft at my full-time disposal and stationed it at Boston's Logan Airport. Commuting between Boston and Cincinnati in an hour and forty minutes was now easy; so was my having offices and secretaries in both locations. Staff meetings were held alternately in Massachusetts and Ohio.

The I.O.I. information system continued, and was now even more vital to our smooth operation than before. (It certainly beats secret or open tape-recording!) My new staff and I instituted an annual three-day TTT (time to think) get-together. Our agenda specifically excluded day-to-day worry items and was limited, instead, to reviews of long-range proposals by our division's strategic planning unit, whose job it was to estimate commercial and military requirements for five, ten and twenty years ahead, beginning in 1962. (The last of the TTT meetings I attended, before retirement in 1980, forecast market needs through the year 2000.) It was clear to each of us that the future would *not* turn out exactly, or even approximately, as predicted by our experts. Nevertheless, we needed *some* long-range scenario of peace, war, mobilization, economic depression, of increase or decrease in business, military and commercial flying. The planners had to guess whether or not some airframe company or group of investors would build a gigantic aerial freighter that could fly, for example, nonstop from central

Brazil to the markets of Chicago in half a day, its cargo hold full of nearly ripe bananas, rather than ship unripe bananas by boats and trains—a trip lasting over one month. Would large helicopters, twenty years from the date of our first TTT session, be the solution for our traffic jams? How many seats would there be in a typical commercial jet transport twenty years hence, and what would be their expected load factor? What would limit the length of a futuristic plane in order to maneuver between existing loading ramps at LaGuardia Airport without colliding with the tail of a plane parked at the opposite gate? Would nonscheduled airlines survive? Would regular airlines survive? Would nonbusiness people pay the large premium to cover the costs for supersonic flights from Paris or London to New York or Washington? There were a hundred more challenging questions. . . . As long as my staff and I kept our feet firmly on the ground and realized that nobody could possibly know the answers, we were safe in not believing the optimistic, glowing forecasts our own strategic planners presented. There was just so much gold in Fort Knox. We at GE were firmly convinced that our engines were superior and our product support the best in the world. Our quite capable—although overconfident—competition, which had powered over 90 percent of the Free World's commercial aircraft only sixteen years ago, certainly would not concur with the prediction that GE would obtain 110 percent of all available business ten years hence. Actually, General Electric's jet engines now power over 60 percent of the Free World's wide-body transports.

The board of directors elected me a corporate vice president in 1963. We were expanding our manufacturing facilities in different parts of the States. GE service shops to overhaul military jet engines were established in California, Washington and Kansas; sales offices in Washington, Dayton, Paris, Geneva, Bonn, Rome, Athens, Beirut, Teheran, Singapore and Tokyo. My staff increased from six to twelve members because I considered finance, legal, employee relations, quality, marine and industrial business, and my S.O.B. critic—while not quite equal in importance to the aircraft engine product departments themselves—vital enough to deserve staff status. I needed their managers' direct input of concerns and their own recommendations.

Although I stressed an open-door policy—i.e., access to me by all nonunion personnel (union contracts forbid direct discussions concerning complaints between union members and management) who had unresolved gripes—I suggested to them that they first visit one of the two "ombudswomen" I had set up in the Ohio and Lynn plants. *Wall Street Journal* took notice on its front page of this type of job, new for American industry, and described the typical responsibility of the two ombudspersons. Their job was to listen to personal complaints, quietly to track down the facts and try to settle the issues satisfactorily to all concerned. The idea for establishing such a position had come to me during the war in China: Our squadron doctor had doubled as chaplain and "punched our ticket" if one of us soldiers had a bitch about which he wanted to talk to someone in private.

To get a personal feel of how GE's large and small military engines were performing in service, I decided to take flights in J79 and J85 jet-powered Air Force planes. Before getting permission to do so, however, I had to pass the altitude chamber decompression test in Dayton, Ohio. While oxygen was being pumped out of the test chamber the testee had to be able to reach a certain altitude before blacking out or losing the ability to judge whether to stick wooden round pegs into square or round holes. I passed the test, then went to Edwards Air Force Base in California for an exciting day: first, a flight in a two-seater Lockheed F-104 "Starfighter," world-record holder in speed and altitude. After zooming along at 1,200 miles per hour in the "missile with a man in it"—48,000 feet above the ground—the pilot of the Starfighter, Captain Bill McCurdy, pushed the stick forward and the plane on its nose. We went into a vertical dive. Over the helmeted head of the pilot in front of me I saw the earth below spinning rapidly closer. At last, McCurdy pulled the airplane control stick back sharply and pushed the throttle of the General Electric J79 engine to full power. My arms weighed a ton in my lap; it was impossible even to lift a hand because of the centrifugal force as the plane's nose rotated from vertical to horizontal. We whipped across California's Death Valley at supersonic speed, narrowly missing the tips of the sandy hills. It was strangely quiet in the cockpit: Moving faster than sound (i.e., flying supersonically), we were leaving the engine noise behind us. I had to look twice at the altimeter before I believed that I saw correctly: Yes, there *was* a minus sign before the digits on the altitude instrument. Captain McCurdy's voice came over my helmet intercom: "Congratulations, Mr. Neumann. I bet you're the first executive to fly supersonically *below* sea level!"

Two hours and one sandwich later, I again donned a parachute and was strapped into the seat of another General Electric-powered twin-jet plane, the new Freedom-Fighter. Northrop's test pilot and I rolled down the runway for my second supersonic flight on that same day in early 1960, this time over George Air Force Base in California. Forty minutes later we were back. I sat exhausted in the cockpit, barely strong enough to open the canopy and breathe fresh air. What was Herman the German doing once again in a fighter plane, seventeen years after his last fighter test flight in China—but flying 900 miles an hour faster?

Before his death, President Kennedy had directed this country to study designs for a supersonic passenger aircraft which would cruise 50 percent faster than the Anglo-French Concorde (which today travels safely in three and a half hours between Paris and Washington, at 1,200 miles per hour). Several United States aircraft companies, our engine competitor Pratt & Whitney and we were in the race for this potential multibillion-dollar business. I organized a special SST (supersonic transport) project reporting to me directly, and appointed as its general manager Ed Hood. (In 1980, Hood was appointed one of General Electric's two vice chairmen.) He, his people and I held design reviews and made sales pitches to the ten major American and many foreign airlines who were potential buyers of such fast planes. From reading publications by our competition describing their engine, I was confident that we had a far superior and much simpler product. Our government planned to make the final selections of plane and engine based on input from its own and the airlines' analysts in the United States. GE's background was superior: The North American B-70 bomber with its six GE engines had flown trouble-free at nearly three and a half times the speed of sound—meaning about 2,400 miles per hour. More importantly, by far the greatest number of supersonic Air Force, Navy and Marine fighters and bombers in service were powered by GE engines. There were thousands of them!

The government experts were well aware that GE had a wide lead in experience with supersonic propulsion. So were the airlines. But I was bothered by reports about the cool reception our salespeople had received at Northwest Airlines. Hood and I flew to Minnesota to meet with NW's chief executive, Don Nyrop. After thirty minutes of the most enthusiastic pitch I was able to muster but during which he pretended to be awfully bored and sleepy, I stopped my presentation and interrupted his yawning: "What's wrong with General Electric?"

Nyrop replied, pointing at the ceiling, "Nothing! Whenever I want a light bulb, I pick GE!" (Later, our lamp salesmen reported that it was *not* GE lamps lighting up Northwest's hangars!) When the favorable voting results came in, we were not surprised: Of ten United States airlines eight voted for GE, one was neutral and only Northwest Airlines voted for our competitor Pratt & Whitney. In 1967, *Fortune* magazine covered GE's determined efforts to win the SST in a six-page feature story. Had Congress not discontinued the SST program by a mere six votes, the American supersonic transport would be alive today and General Electric engines would propel the winning Boeing plane! (The operational results of the English-French Concorde supersonic program show it to be a great technical success—and a financial fiasco.)

One day I visited Daimler-Benz and their automobile museum at Stuttgart in West Germany, where was exhibited one car of each model they had ever built, from the very first Benz to the latest Mercedes diesel. This gave me the idea of setting up a GE jet engine museum in our Ohio plant: In the decades to come, it would be educational and fascinating for future generations of engineers. They would be inspired by the incredible progress made since this country's first jet engine was run by GE in 1942, by our designs of faster, lighter and smaller engines, with more power for each pound of their weight and more thrust for each pound of air flowing through the engine. Tremendous advances have also been made in the development of new raw materials, modern technology of chemical milling, friction welding, laser-beam drilling, powder metallurgy, plastics, X-raying of running engines, computer-controlled machine tools and fabulous inspection methods through ultrasonics and bent-light beams.

During the early sixties our advance design engineers had several bright ideas regarding the possibility of extracting still more thrust out of a jet engine power plant. One proposal that looked most likely to succeed was a giant fan (a compromise between a three- or four-bladed airplane propeller and a sixty- to seventy-bladed compressor stage) mounted in front of the engine's air inlet. The total air mass moved aft by this fan (which was rotated by a shaft coming from the engine's turbine) was split into two parts: about one-eighth of the air was sucked into the conventional engine inlet, compressed, mixed with fuel and heated; the other remaining, much larger amount of air, about seven-eighths of the total, was pushed aft through a rear nozzle bypassing the engine itself, neither burned nor flowing through its turbine. This "high-bypass airflow" could give an engine lots of extra thrust especially during a plane's takeoff and climb, and would cause lower fuel consumption during cruise. My staff and I carefully studied the various engineering ideas of this novel type of engine. At that time such a system of propulsion was possible only because of GE's leadership in turbine technology, specifically in the technically advanced internal air cooling of the white-hot turbine blades. Calculating the fuel consumption for each pound of thrust of a high-bypass engine, we came to the startling conclusion that it would save 30 to 35 percent of fuel vis-à-vis the conventional jet engine used by air forces and airlines at that time. (In "real life," there would be some penalties in weight, cost and aerodynamic drag because of the larger size of the front fan and its casing.) Combining the pluses and minuses, we arrived at the still incredible net savings of about 25 percent in fuel consumption.

Finally, I was satisfied enough with the design details to have a demonstrator engine built with high-bypass and variable stator features. One year later we had test results, which were phoned to me while I was in Buenos Aires: They were spectacular! I returned immediately to review the sales strategy proposed by my staff. During one of our rare Sunday afternoon sessions, my key technical people and I made one more final effort to find out what possibly could be wrong with the favorable conclusions reached, which sounded too good to be true. Regardless of how hard we tried, every detail withstood any challenge from us reviewers. After dinner that Sunday evening, I phoned the head of the Air Force's research and development division, Major General Marv Demler, at his home near Washington, D.C., and asked him for an audience the following day "in strictest privacy."

Early next morning, behind closed doors in the general's office,

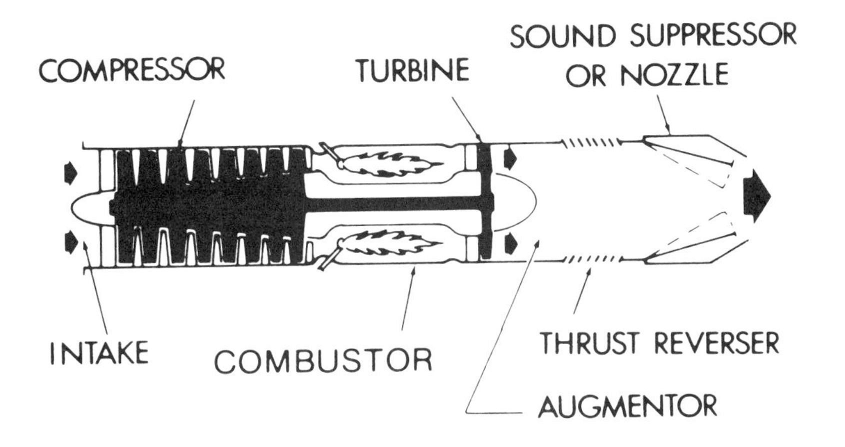

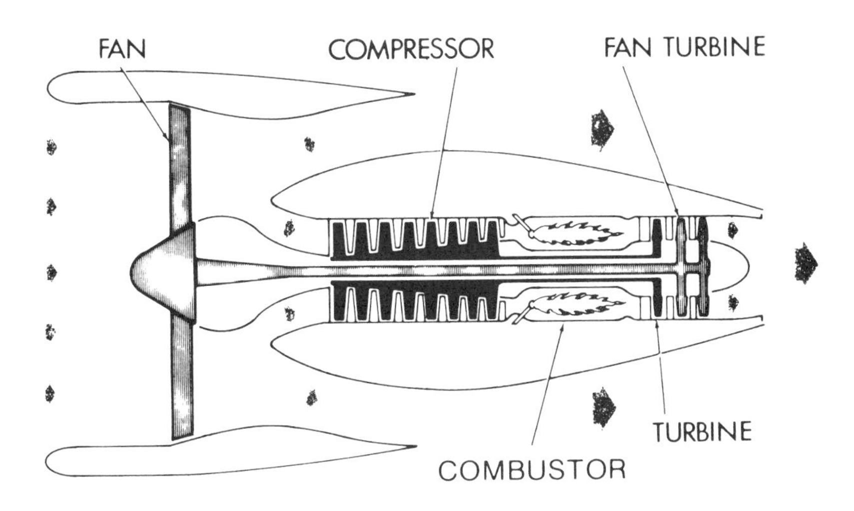

I unrolled a cross-sectional drawing of the new type of high-bypass engine we had just run. The drawing stretched from wall to wall of his office. The general was stunned and pleased at the technical breakthrough, to say the least; with my concurrence he briefed, over the phone, German-born four-star General Benny Schriever, who headed all our Air Force's advanced programs including missiles. The advent of such a high-bypass engine, with its very high turbine temperatures, more than twice as powerful as the largest engines then flying, could well revolutionize air transportation. (It did.) The range of new transports using such engines could be dramatically increased. Construction of huge planes that could transport the heaviest Army tanks, fighter planes and helicopters or over seven hundred fully equipped troops in one single plane would now be possible. If a second deck were installed, the plane could carry over one thousand combat troops over a long distance. A brief mandatory competition amongst airframe builders and engine producers was launched immediately. Lockheed won the contract for the world's largest airplane, the Galaxy C-5A, and General Electric—with its two-year head start—won the production engine contract as we had expected. The development contract was for $459 million, the largest single contract GE had ever received by then. In addition, the engine noise turned out to be substantially quieter for that type of jet than for the then-current generation of engines. Three and a half years later I sat on the dais with President Lyndon Johnson and other dignitaries in Marietta, Georgia, as the first giant C-5A plane with four GE high-bypass engines hanging under its wings was towed out of an enormous hangar for its first public showing.

Boeing's 747 "Jumbo" jet, quite a bit smaller but faster than Lockheed's C-5A, is a good example of what was done with engines of the new high-bypass type. While GE was busy to the bursting point doing the work for the Air Force's C-5A, had also won a nearly half-billion-dollar contract for the development of America's supersonic transport (SST) engine, and at the same time was working on a competition for the engine of a low-flying, supersonic and long-range American B-1 bomber (which GE also won a few months later), Pratt & Whitney had lots of time and was free to develop a high-bypass power plant sized and tuned specifically for Boeing's 747. Two years later, I felt that we had adequate manpower and company funds available to modify our military engine for the C-5A into the commercially suitable CF6 power plant for the Douglas DC-10, the European (primarily French and German) Airbus A-300 and finally, a few years later, for Boeing's latest 747. This gave airlines a choice of engines, and Pratt & Whitney some competition!

As part of my "walking the shop" philosophy, I spent several weeks in 1967 and 1969 at the Vietnamese war scene, following the granting of my request by the Vice Chief of the Air Force, later Commanding General of the Strategic Air Command, four-star

General Bruce K. Holloway. (He had been my fighter group commander in China back in 1942.) I sought to get direct input about the performance and maintenance of GE's military engines in Viet-nam's hot, humid and very dusty climate. And to find out—by talking to my own thirty-two technical representatives living with American and Vietnamese squadrons there, also by interviewing Air Force, Navy and Marine pilots and mechanics—whether there was any way in which we at the home front could do a better job and give more effective assistance to our men in southeast Asia. I visited the air bases from Bien Hoa near Saigon in the south, all the way to the tip of the northern front, where our rescue helicopters operated picking up bailed-out American pilots from behind enemy lines. I spent a couple of days on the nuclear aircraft carrier *Enterprise,* which was cruising in the Gulf of Tonkin, to observe maintenance of GE engines for fighters and 'copters performed on the high seas. Whenever I traveled in Viet Nam, it was by helicopter, two-seater fighter plane or Ambassador Henry Cabot Lodge's jet.

Once, I visited with a Marine rescue detachment in the north (all Marine helicopters use GE's engines exclusively). I arrived late in the evening by chopper and slept in the commander's tent. Because of frequent rocket attacks, each of us covered himself with an armor-plated flak vest. Next morning I had breakfast with the men. One of the gray-headed Marines turned out to be a GE technical representative of my division. When I asked him how long he had been in Viet Nam and in such an exposed location, he replied, "Four years." I apologized to him for my management's oversight and offered to have him transferred back to the States at once. "Please don't" was the tech rep's plea. "It's wonderful here. Besides, I can't stand my wife. . . ." he added jokingly. A local girl was preparing his meals and washing his laundry.

After each return to the States, I briefed top brass in the Pentagon, warned them repeatedly that the engine-support phase of the Vietnamese conflict was not necessarily indicative of what might take place in another war. For example: Our side was being neither bombed nor strafed, and therefore planes and engines were rarely damaged on the ground. We had unlimited time to send engines for overhaul all the way to the U.S., Japan or the Philippines. After my return to the States, I sat with my own project managers and writers of technical maintenance manuals and told them how their important paper work, written in the peace and quiet of an air-conditioned office, was often too complicated to be understood by our GI's in Viet Nam.

While near the end of my second visit in Viet Nam, I received a cable from my very capable secretary, Shirley Clarke: "PALACE REVOLUTION. RETURN IMMEDIATELY." I knew that she wouldn't kid about so serious a matter. I took the direct flight from Saigon via Tokyo and Chicago to Boston. Here is what had happened: Two days after I had left the States for Viet Nam, anonymous letters mailed in Cincinnati were received by high Defense Department officials, GE's top management and by each of our directors. Some of the letters had been misspelled, some were handwritten, most of them typed. Each letter accused me of improper business conduct. During the investigation of who wrote such letters, we had the fullest cooperation of the Defense Department. Working at night, a specialist typewriter detective from Chicago identified two of the three machines used, where they had been purchased and where they were located in the Evendale plant. Shirley and my brilliant legal chief counsel, Jim Sack—both crossword puzzle fans, using their imagination—solved the mystery of whodunit. They and the detective put together all available bits of facts that identified the culprit. Much of the incriminating evidence came from one of the accusing letters, which had been typed on paper used by the writer as a backup sheet sometime before. The barely visible colorless imprints of a part of the original letter's contents revealed the writer. For some very personal reasons, the secretary of one of my staff members wanted to see her boss take my place (he claimed not to have been aware of her letter campaign). The secretary was fired promptly; her boss resigned shortly thereafter.

In 1968, GE expanded its organization to keep up with the increase of business. Ten groups were formed instead of the existing five. Our Flight Propulsion Division was elevated to the status of Aircraft Engine Group (AEG). I became its group executive and my staff moved from department to division levels. This was great on paper, in privileges and in our paychecks—yet our fundamental responsibilities remained the same as they had been since 1961. By 1968, the size of my group, worldwide, exceeded 31,000 employees. Sales-wise, as our commercial engine business shifted into high gear, our group produced about 10 percent of total company sales. With the invaluable help of airline friends in the United States and in Europe, such as American and United, Lufthansa, Swissair, KLM, Air France and Alitalia, as well as South America's Varig, the Orient's Thai Airways, Japan's Toa Domestic Airline and many more who clearly indicated a preference for commercial jet engines made by GE, we established a worldwide customer service net for their planes powered by our high-bypass CF6 jets. At the same time, Rolls-Royce had signed an agreement with Lockheed Aircraft to be the sole supplier of engines for their new airplane, the L-1011. It was clear to me that Rolls, which did not yet have the technical background for high-bypass engines that GE and Pratt & Whitney had, were overcommitting themselves badly in time and money in their hasty effort to catch up with us. Their claims were so incredible that I suggested to our board of directors that "I should be fired if Rolls comes even near to what they promised Lockheed and their airline customers."

This is the way it turned out: I was waiting to board a plane at Rome's Leonardo da Vinci Airport when an Alitalia executive spotted me and came rushing over, waving an Italian newspaper with big black headlines and shouting an incredulous *"Mama mia!"* He then interpreted the paper excitedly: Rolls-Royce had declared bankruptcy! The British government was considering the takeover of the Rolls-Royce aircraft engine works at Derby. What had completely puzzled me was the naïveté of Rolls's presumably sophisticated customers who had believed Rolls' promises and who certainly could have, or should have, known better. Their judgment was obviously impaired by the glamorous Rolls-Royce name and the leather-bound propaganda material issued by the Derby outfit.

It was most gratifying that nearly all Europeans—in addition to five major United States airlines—operating the new DC-10's selected GE's engines over Pratt & Whitney's and Rolls-Royce's. Our share of the U.S. and international commercial market grew by leaps and bounds. Today, 1984, over one-half of all wide-bodied transports operating and ordered in the Free World are powered by General Electric engines, including most of the French-German European Airbus A-300's and a great number of Boeing 747's. Seventy-five airlines are using GE's engines.

Not all negotiations for multimillion-dollar commercial deals took place in the plush offices of airlines or of airplane builders. I reached a very important handshake agreement with Boeing's chief executive, T. Wilson, in his small room in uptown New York in August 1973. His room had only one chair, so I sat on Wilson's bed. He welcomed GE's decision to enter the engine competition for his 747, now that we had some extra capacity due to the cancellation of the SST by Congress. Wilson promised to give henceforth any buyer of Jumbo jets a choice of engines. We had been friends for many years, and had developed a mutual respect for each other's word. Within a couple of months, Boeing/GE made the first sale of a number of 747's powered by GE to serve as airborne command posts for the President of the USA in time of war. In quick succession followed orders for GE-powered Boeing 747's from key European airlines, which had been pleased with GE's fine support and the satisfactory operation of its engines on their DC-10's.

Overseas, Germany was one of my favorite countries to work with. At Lufthansa we scored a major coup against our competition when this highly respected airline, under its technical leaders Hoeltje and Abraham, replaced each of its Pratt & Whitney-powered Jumbo planes with the latest Boeing 747's powered by General Electric engines. The same engine had been selected by Lufthansa for its DC-10's and A-300 Airbuses. Following the lead of Lufthansa, many large and small airlines around the world that could not afford such expert technical staff as was employed by the Germans specified General Electric CF6 engines.

Over the years I had interesting contacts with the British Ministers of Transportation. One of them told me in London that he would always select a Rolls-Royce engine for the British government-owned airline "even if their engine were technically inferior." He asserted that it was cheaper for England to support Rolls-Royce workers in Derby (where their aircraft engines are made) than to pay them unemployment compensation. Interestingly, the three British charter airlines that operated wide-bodied aircraft—and which were not under British government control but free to select their own engines—chose General Electric engines over Rolls-Royce's.

I was deeply disappointed to learn of two cases in which our own supposedly neutral government tried (and in one of the two cases succeeded in doing so) to swing the selection of aircraft and engines to certain manufacturers. I happened to be in the office of one U.S. airline top executive when he received a call from a high government official in Washington who put the pressure on him to purchase aircraft A over B. A was using the Rolls-Royce engine. Thanks to that airline's strong-willed president, it went ahead with its own choice of aircraft and selected plane B with GE's engines. In the other, even more disgusting case, I was told by a very high-ranking American official who was an eyewitness to that meeting that a President of the United States accompanied by his Secretary of State (while both were in office)—in a meeting outside the continental limits of the USA—"suggested" to a foreign prime minister that the government-supported airline purchase aircraft A with Rolls-Royce engines. One month later, I received confirmation from an independent source that the report was indeed true.

As part of GE's engine business going international, we licensed the manufacture of several of our engines in foreign countries (including Japan, Germany, Israel, Italy, Belgium, France, Great Britain, Canada and Taiwan) with the full concurrence of six departments of the U.S. government, each of which could have vetoed any one agreement. Even Soviet Russia and Communist China wanted to license some of our commercial engines, but our government disapproved. Of course, we cannot prevent other countries from selling GE spare engines to Russia or China. We also established sixteen service shops worldwide—from New Zealand to Greece, from Singapore to Paris—employing over three hundred technical representatives.

In 1971, we participated in a battle in which our two competitors, Rolls-Royce and Pratt & Whitney, were trying to become a partner in a joint venture with the French government-supported engine manufacturer SNECMA, to develop their first commercial engine. The president and general manager of SNECMA was René Ravaud, a former naval designer and *Ingénieur Général* of the French Army. Ravaud, a big, tough soldier and a charming person, had lost his right arm in France shortly after D-Day 1944, during the battle for the liberation of Brest harbor. There could not be a more courageous, intelligent and enthusiastic leader; Ravaud was much admired by French presidents and members of their cabinets. He and I clicked from the very first moment we met, and General Electric was formally selected to be France's engine partner.

We agreed to make a real partnership out of our joint efforts, and to split the responsibility evenly between our two companies: Each was to design, develop and produce one-half of a new, medium-sized high-bypass jet engine, to be called the CFM56, with a 50/50 (rather than the usual 51/49) percent share of responsibility. We created a small management company, CFM International, registered both in France and in the United States. (We Americans silently blessed the Lord that He made the French so clever that they could conduct all business with us in English.) This new engine, the CFM56, is on the way to becoming a great international success. French engineers have always been known to be ingenious; in this CFM56 project they were especially imaginative, cooperative, efficient and careful. Ravaud's sense of Gallic humor was unsurpassed: I once sent him a cable to Paris that we had succeeded in a certain negotiation "even if it cost us an arm and a leg." Too late I realized what message I had sent to one-armed Ravaud, and made every effort to retrieve this cable. But it was too late. I received a cable next morning: "IT IS BETTER TO LOSE ONE'S ARM THAN ONE'S HEAD. RAVAUD."

GE was formally threatened by the Arab League Boycott Committee for providing engines and support to Israel. Not, as we expected, for any military help we gave her, but because we sold Israel engines for their new business jets, built near Tel Aviv and sold the world over. Our chief executive at that time, Fred Borch, instructed me to respond to the two representatives of the Arab League Boycott Committee who met with me at the Hotel Pierre in New York: "If you want to buy anything from General Electric, we will be most pleased to sell it to you. If you do not want to buy from us, that is your own decision." No Arab boycott ever followed as a consequence of GE's engine deals with Israel.

When the Arab oil boycott in 1973 was enforced nationwide, it gave me the justification I had been waiting for since I was a student in Germany: to own, once again, a motorcycle. I bought a powerful BMW, a lifelong dream of mine. When I rode it to work the first day, a guard at one of the entrance gates to the Lynn plant stopped me. (The plant guards had long-standing orders to let neither motorcycles nor bicycles enter—for what reason no one knew.) The officer at the gate had not recognized me through the visor of my helmet. Only after he discovered whom he was preventing from entering was I let into the plant. It took three hours before the official word was issued: Henceforth, motorcycles and bicycles were welcome! (Five years later, 650 cycles got entrance permits.)

Since I tried to talk to as many people from different functions and different organizational levels as possible, I made it a habit to eat in the cafeteria whenever I was at the Evendale plant. One day a factory foreman stood in the lunch line directly in front of me. I asked him about his activities and ambitions. This foreman in blue coveralls, Bud Bonner, invited me to visit his shop area during one of my nightly walks through the plant. (I did show up late one evening and noticed the exceptional neatness and orderliness that existed in his area.) He also bragged that he and his workers could do any manufacturing job in half the time other organizations in our plant could. It so happened that we needed several parts for the J79 engine in a great hurry. My "regular" people gave me a "best time" estimate of nearly two weeks. I bypassed organizational channels and asked Bonner how long it would take him to make the parts. His reply: six days.

I gave Bonner the job, settled four complaints by the unions, two by my own management—and had the needed parts available in five and a half days! As a consequence, I gave Bonner one of our field service shops to manage, and kept on promoting him in recognition of his continued demonstrated outstanding performance. Ultimately, he became the general manager of our rapidly expanding marine and industrial business. Bonner won 100 percent of the U.S. Navy's orders for their new gas-turbine-powered ships with engines derived from aircraft jets. All new U.S. destroyers, frigates and hydrofoils are today powered by GE. Bonner also inspired Germany to propel her Navy frigates with our group's marine engines; other NATO and South American countries were similarly motivated. Bonner is truly self-made; he never finished high school because he was brought up in an orphanage that could not support him beyond the age of sixteen. He is intelligent, enthusiastic, energetic, a good salesman and a leader. In 1978, I recommended Bonner's promotion to corporate vice president, but was doubtful of the outcome: To promote someone who was not a college graduate to such an elevated position had rarely before happened at GE. Our personnel people felt strongly that a degree from a reputable institution of learning was a minimum requirement for someone to become an officer of the General Electric Company. Vice Chairman Jack Parker and Chief Executive Reginald Jones, however, properly made "demonstrated performance" the main criterion. Today, V.P. Bud Bonner's success is a tribute to the intelligent, flexible attitude of GE's chief executives, its fine board of directors, and a democratic America! "We need more men like Bud Bonner!" our retired chief executive wrote me the other month.

Traveling a few days every week for sixteen years to visit with military and commercial customers became very wearing. Latin America, from Argentina to Mexico; the Orient, from Korea to New Zealand; Europe, from Spain to Finland; Africa, from Tunisia to the former Belgian Congo—not to mention the many military and airline headquarters in the USA—all had to be visited by the boss man, or customers felt neglected. Flying at night, conducting business during the day, trying to be home over the weekend to see the family, reading office mail brought to the airport in packages by the chauffeur and leaving instructions for my secretary because I was back in the air again Sunday night became a 110 percent activity that left me no time to read a nontechnical book, a weekly magazine or even newspapers. It was push, push, push all the time —except for my vacation, which I split in half: sailing my 38-foot ketch in summer, and skiing in winter. Even those vacations were sometimes interrupted: Once a Coast Guard helicopter located me under sail outside Nantucket Island in the Atlantic and told me to call my office regarding a government request; a U.S. Army tug stopped me in Cape Cod Canal to deliver an urgent message; a phone call from the States woke me past midnight in the French ski resort of Tignes asking me to return to Washington immediately. I also *had* to find time, somehow, to walk the shop and attend staff meetings. Damage to my health by this hectic pressure was unavoidable: I had never missed a single day of work in twenty-nine years; but suddenly, in 1977, I needed a major heart overhaul, with bypasses and a plastic heart valve.

Eight weeks later I was back on my motorcycle; but one year after open heart surgery, the doctor urged me to slow down, to sell my BMW and to quit skiing downhill. I informed my bosses, Vice Chairman Jack Parker and Chief Executive Reg Jones, in 1978 that

I should retire because I would be unable to continue at the tempo and with the energy I considered essential to keep the fast-moving AEG train rolling. Besides, with my staff and myself running the business for seventeen years, it was likely that we had developed blind spots in our operation, that we were overlooking and perhaps missing some good opportunities. A change in management would be healthy. I nominated as my successor hardworking, clever and loyal Fred MacFee, who had been my right arm for many years. The transfer of the reins to him proceeded without a hitch. But MacFee was near retirement age himself. A former British aerospace engineer, Brian Rowe, took over the jet engine operation from MacFee after one and a half years. Why, one must ask himself, did these two "damned foreigners," amongst the thousands of Americans already aboard the Aircraft Engine Group, get these key jobs? Both of us had the advantage of years of tough apprenticeship before being admitted to engineering schools; both of us were hard-boiled and practical.

During the last fifteen years of work at GE, I changed neither my office nor its furniture. Behind my desk hung a large sign: FEEL INSECURE. Below it were four framed sayings which meant a lot to me. One read:

> Duty as Seen by Lincoln
>
> If I were to try to read, much less answer, all the attacks made upon me, this shop might as well be closed for any other business. I do the very best I know how—the very best I can; and I mean on keeping doing so until the end. If the end brings me out all right, what is said against me won't amount to anything. If the end brings me out wrong, ten angels swearing that I was right would make no difference.

The three smaller framed signs read:

> Show me a good loser and I'll show you a loser.
>
> Nobody is completely useless. He can always serve as a bad example.
>
> The harder I work, the luckier I get.

Those represented the philosophy by which I guided our business. I firmly believe in each of them. I have been praised much more than I honestly deserve, since what I am generally credited with is truly the work of hundreds and thousands of team players who produced the detailed designs and managed the production programs, who came through in peace and war, on schedule, with the expected reliability and performance.

I was a co-recipient of the Collier Trophy in 1958, received in 1970 the Goddard Gold Medal, was given the International Guggenheim Award in 1979. The Evendale suburb of Cincinnati named the road leading to the big jet engine plant off Interstate Highway 75 Neumann Way. Years ago, I became an honorary member of the faculty of the Industrial College of the Armed Forces, whose students I have addressed annually since 1961. In 1977, Valéry Giscard d'Estaing, then President of France, awarded me the highest decoration a foreigner can get from his country, the Ordre national de la Légion d'honneur.

When I retired on January 1, 1980 (the same day Jack Parker did), I closed the book on thirty-two years of exciting, fully satisfying work with the General Electric Company. I have nothing but the highest of praise for the fair, opportunity-giving, inspiring leadership of this great company. Of the many awards and farewell presents received, I appreciated one more than any other: a thick, leather-bound book embossed with the words "THANKS, GERHARD NEUMANN," containing more than ten thousand signatures from the Aircraft Engine Group's workers and managers.

I was particularly pleased when the Governor of Ohio declared a "Neumann Day" and the union leadership gave me a "management-free" banquet attended by their people exclusively, with speeches by seven union presidents. "There are times when we are all on the same side of the fence, and one of these times is tonight," one union president said. My Lynn associates gave me a surprise dinner with one thousand management people and union representatives; Anna Chennault, widow of my commanding general of the Flying Tigers, was the main guest speaker during the "Gerhard Neumann Night." The Lynn Chamber of Commerce gave me a breakfast, with businessmen, the speaker of the Massachusetts House and several former mayors attending. At the annual Golden Gate Award dinner in 1981 (an award for foreign-born men or women who make "most important contributions to the United States") Massachusetts Governor Edward J. King, SNECMA president René Ravaud, and Northrop's chief executive, Tom Jones, honored me with their presence. I appreciate all these signs of recognition and many more. But *the fact is* that it was the thousands of workers, engineers and support people, most of whom spent their years loyally and anonymously, that made General Electric's jet engine business grow successfully to the fine state in which it is today.

Yes, there was a lot of satisfaction and pleasure in being one of the many who made General Electric's jet engines a major contribution to national and NATO defense and to the flying public. Our progress was dramatic. The support from the management of the company and its board of directors was superb. From close relationships with our customers over many years, I learned one thing: Building personal confidence, listening to what the customer thinks—regardless of what the computer says—keeping my promises and an all-out, stubborn dedication to meet or exceed commitments that I made were the most important reasons for success in my job. An occasional engine failure was forgivable—even expected—but failing to make every effort to fix problems and to meet promises was not.

19

Wenn Man Reist, Dann Kann Man Was Erzählen (When One Travels, Then One Can Tell Stories)

German tourists are known to be the world's most ubiquitous travelers. I probably rank among the top 1 percent. My journeys over the past forty-four years were principally dictated by work and by wars. I circled the globe several times in either direction during the last thirty years, and flew half around the earth many times in north–south and east–west directions. There is hardly a civilized area in this world where General Electric's military and commercial sales and service of its jet engines did not require me to go. This self-imposed requirement was part of my philosophy of walking the shop. Wherever I traveled, it was by air. In matters not pertaining to General Electric, I went by troopship to the States; by Jeep across Asia; by weapons carrier during a private photo safari in Kenya and Tanzania with my family; and in my 38-foot sail ketch during a sailing vacation to the Bahamas and back to Boston.

To save time on overseas trips, I flew mostly during the night and began work the following morning. I shaved and changed my shirt aboard and was ready for customer sessions and plant visits from the moment the plane landed until I left late at night. The trips were scheduled so tightly that I rarely saw anything of the countries I visited other than the inside of the cars of my local GE technical representatives who met me at the airport. During the drive to town I was briefed on the good and bad I was going to hear from the military or commercial customers, and what responses or commitments the reps suggested I give. They also reported about the responsiveness (good or bad) of my people at home as viewed by those stationed far away. I considered my service representatives the most knowledgeable source of information about the *real* performance of our products in the hands of actual users. To avoid any possible retribution for their "leaks to Mr. Neumann" at the hands of my project managers at home, I established a separate Product Support operation reporting to me directly.

Besides our various travels that led us all over the globe to the common and not-so-common places mentioned in *National Geographic* magazine, Clarice and I had some experiences worth reporting here.

On August 31, 1951, we set off on a visit to Acapulco, on the Pacific coast of Mexico. Luckily for us, we were late at the airport and were assigned two seats all the way in the back of a brand-new Douglas DC-6B, belonging to Compañía Mexicana de Aviación. Acapulco was just beginning to develop into an international playground. The only way to get there in those years was to fly to Mexico City, then drive over primitive roads to the Pacific coast. Our plane schedule from Los Angeles called for arrival at Mexico City slightly before six in the morning. Six miles from Mexico City's airport, above the "dry lake" Texcoco, which fills with water six to eight feet deep during the August to October rainy season, we saw faint daylight coming over the horizon.

The "Fasten Safety Belts" sign lit up. We descended from cruising altitude in a steep spiral. Our DC-6 was just entering what must have seemed to our pilot a layer of clouds when—accompanied by an ear-splitting crash—the right wing tip of our plane sliced into the surface of the lake. The enormous momentum of a big aircraft plowing into the water at 200 miles per hour ripped off the right wing together with engines and propellers. The one-winged fuselage cartwheeled over the surface of the water and tore open from top to bottom at several locations between cockpit and tail. The plane's belly was smashed as the DC-6 bounced twice into the air, breaking into segments, then settled gurgling to the bottom of the lake, supported by whatever remained of its extended landing gear. For a moment I simply couldn't believe that we were still alive.

There was dead silence for about ten seconds, then began the moaning of the passengers. Lake water topped by gasoline reached up to our stomachs. Seats had torn off the floor; cabin side and ceiling panels had collapsed. Emergency lights did not work. I recall an open box of candy and someone's jacket floating by my chest; Clarice remembers knives and forks flying through the air like missiles during the series of impacts, apparently off breakfast trays. It was a disaster that became more and more visible as daylight began to illuminate the eerie scene. Our two seats were still attached to the plane's floor, and the safety belts had held us. I urged Clarice to hurry and reach a two-foot-wide jagged crack in the left side of the fuselage ten feet ahead of us, and to twist herself past the rough edges onto the left wing, whose tip was sticking out two feet from the lake. She managed—barefooted—and I followed her. We were joined by three other shoeless travelers (who lost their footwear in the impact), including a Pan Am stewardess on vacation. My main fear at this very moment was the typical post-crash gasoline fire, usually caused by fuel self-igniting after it touches hot engine components. I had seen this many times during the war. The two propellers of the left wing, still attached to the fuselage, had sheared off their engine shafts and sliced through the aluminum fuselage into the side of the occupied front section of our DC-6. The fact that all of us were not killed instantaneously was due to the piecemeal destruction of the plane, in a series of three separate impacts rather than in one single head-on crash, and that miraculously no fire started.

A few minutes later, Mexican Indian fishermen appeared like ghosts out of the mist, standing wrapped in serapes in their *chalupas* (hollowed tree trunks like those used as canoes by our American Indians). None of the natives made any effort to come closer than 150 feet to the wreck, which looked more like a whale lying on its side with its guts torn out than a flying machine. One of the fishermen rolled himself a smoke and took a match out of his pocket to light his cigarette. Only the yelling of the vacationing Pan Am hostess, who knew some of the local Indian dialect, saved us from getting roasted by the inevitable gasoline fire.

The Pan Am hostess was able to talk the fishermen into bringing their *chalupas* to the wing tip where we stood shivering. One after the other, we lowered ourselves carefully into five wobbly boats pushed by long oars to the nearest shore, two miles away. Ten minutes later, an Air France DC-6 dropped a rubber life raft over the wreck. Cars and ambulances began to arrive at the water's edge from a nearby road; a Red Cross tent was being set up. Not only were we soaked in water and gasoline, but our clothes had large spots of bloody-looking red dye from the plane's woolen blankets which we had used during the night. Once ashore, we five uninjured passengers were rushed to a luxurious downtown hotel, bypassing all immigration procedures.

The following morning we expected to find big black headlines and photos in Mexico City's newspapers covering the plane crash and to read what happened to the crew and the other passengers . . . but there was not a single word about the accident. Something smells here, we suspected. Whereupon Clarice and I visited the airport manager; his conclusion was the same as mine: pure pilot error. A wrong altimeter reading and mistaking, in the dawn, the layer of mist for clouds, immediately above the water. He told us the reason for the media's dead silence: Mexicana, a government airline, had had another major accident only a few months earlier. The airline hoped to keep this latest crash under wraps, if at all possible, in order not to scare off potential passengers. Whereupon Clarice and I chartered a Piper Cub with a Mexican pilot, circled low over the wreck and took photos. (Back in Massachusetts, after a fun week in Acapulco, I sent some of my prints of the wreck and a brief report to the world's leading aviation magazine, *Aviation Week,* which printed my letter and one photo headlined: "Mysterious Crash in Mexico?" A doctor taped a couple of my cracked ribs, after we got home.)

In June 1967, we had another type of travel excitement. I had to attend the International Paris Air Show, a biannual affair of what's new in aviation. Ten minutes after Clarice and I left Athens for a trip to Israel and Turkey via Greece, our pilot reported that fighting had erupted between Israel and her Arab neighbors. Our TWA flight continued on to Tel Aviv for fifteen more minutes, then the pilot announced a change: We were going to head for Beirut instead. Okay, I thought, it will be interesting to revisit the place where I had been in 1939. Half an hour later, though, another announcement from the cockpit: Fighting had broken out across the whole Mideast, and Lebanon had closed her borders and airport. Our plane would be heading northeast for Teheran instead.

Two days later, an unscheduled Dutch KLM jet freighter with military cargo was leaving Teheran for Tel Aviv in the evening so that it would arrive there at midnight. Clarice and I decided to fly along, curious what would happen next. We arrived at blacked-out Lydda (Tel Aviv's airport) and hitchhiked to Hotel Hilton in KLM's crew car, because taxi drivers were driving heavy Sherman tanks at that moment! Every one of the Hilton's personnel had been drafted except old cooks and a few ancient maids; guests had to make up their own rooms; meals were not being served: One had to help himself to whatever had been set up on a large table in the dining room. When Israelis talked of total mobilization, that's exactly what they had in mind: *total.* Each male between sixteen and fifty-six and every woman from eighteen to thirty-six was called up; unless he or she was dumb, deaf or physically a wreck, he or she would be wearing a uniform within twelve hours. Each privately owned motor vehicle and taxi was requisitioned; milk trucks were converted to fuel tankers to support mechanized equipment in the field. Mail was carried by fourteen-year-old girls wearing khaki and postman hats.

Military bulletins issued at Tel Aviv were incredibly favorable to the Israelis, claiming that Arab air forces had been totally wiped out at their home bases within hours after the fighting started. It turned out to be completely true—except that they were wiped out not *after* the war had started, but in "preemptive" Israeli air strikes that began three hours earlier, we were told. The Sinai desert and Suez Canal had already been crossed by Israeli armored columns when we arrived; over three thousand Russian-made tanks were captured on the Egyptian front alone; war prisoners amounted to tens of thousands. The only real opposition to the well-oiled and precision-trained Israeli war machine came from Jordan's Arab Legion. The next day, Clarice and I got a room in Jerusalem's King David Hotel on its top floor facing east, overlooking the Old City with its Blue Mosque. Near us was the two-thousand-plus-year-old city wall; we filmed from above Israeli sappers dislodging Legionnaires by armored bulldozers turning over occupied pillboxes. I took a movie of a stirring scene when Israelis moved the barbed wire barricades blocking the path to their holy Wailing Wall, from which Jews had been banned since 1948. Two days later the Six-

Day War was over. I congratulated my customers and carried out what I had come to discuss with their Air Force. Then we flew on to Istanbul and the Bosporus.

Who said you can't go home again? It was August 1976: The Communist government had granted my request to let four Neumanns stay twenty-four hours in my former hometown, but had refused permission for us to *drive* from Berlin. We arrived at Frankfurt/Oder by express train from East Berlin. I was eager to show my wife and children—and to see for myself—what was left of my father's feather business and the home where I was born and brought up. We had little expectation to find anything that had belonged to the Neumanns. Postwar German statistics reported that my hometown, Frankfurt an der Oder, was 80 percent destroyed in the last-ditch battle between German reserves and the Russians during March and April of 1945. Deutschland's capital, Berlin—the ultimate prize—lay fifty miles to the west of the Oder River.

Since my last stay in Frankfurt/Oder, in 1939, Nazis, German soldiers, hordes of refugees from East Prussia, Russian and Polish troops bent on bloody revenge, East German Communists—one after the other had moved in and out of the old Hanseatic town founded in 1252. They plundered her factories, tore down monuments and everything historical that could possibly remind anyone of the years 1871–1945, years of the imperial, socialistic, democratic and Nazi regimes. At the time of our visit, World War II had been over for thirty-one years. The new Democratic Republic of Germany (Communist) had barely begun to replace parts of the ruined city with modern, Russian-style housing and one good-looking high-rise building. The rebuilt *Bahnhof* (railroad station) seemed unchanged. One had to walk from the train platform underneath the tracks past the station's toilets to reach the square where taxis used to wait. The identical pungent odor pervaded the underpass in 1976 as it did fifty years ago! But there were no taxis. It was a ten-minute walk downhill from the station to the site where my father's factory used to be. Some houses along the hill were missing. Others had been badly damaged during the war and were only partially repaired. At the time of our visit in 1976, wooden boards still were replacing some shot-out windowpanes.

I found myself walking faster and faster, practically running ahead of my family as we made the last turn onto Gubenerstrasse. Only 500 feet to go! And then . . . For a moment I thought I must have lost my mind. Staring me in the face were the same old iron gate, the same four-story office building, the factory, the powerhouse with its high chimney, the truck and car garages, even the huge old chestnut tree—everything was there as I had left it. And to top it all: Six-foot-high black letters on a white background, painted along the fourth floor of the factory building at least fifty years ago, still read: NORDDEUTSCHE BETTFEDERNFABRIK. SIEGFRIED NEUMANN. I thought I was dreaming.

We then headed for my home in the other part of town. As I remembered, it used to be a half-hour walk from our house to my father's factory—and my nearby Gymnasium—but now all distances seemed to have shrunk dramatically. It was only ten minutes later when we rounded the corner to Humboldtstrasse. Of the original eighteen houses on one side of that narrow, one-way street, more than half had disappeared. The other side of Humboldtstrasse had been an old cemetery with huge trees; it was now a neglected park. As we walked down the street, I spotted—hidden behind trees which used to be much smaller when I had seen them last—our villa. Our #11 was standing tall and handsome as it had for the past sixty-five years. Except for shrapnel impacts on one side, everything else was just like in the old days. I often had told my children, who asked to be driven to school in Swampscott when there were a few inches of snow on the ground, how spoiled Americans were and how we German children always had to *walk* thirty minutes to school, in *any* weather. . . . Remembering this, my children didn't believe me; but my wife saved the day: "Daddy had shorter legs in those days!"

We stopped outside the gate in the house's garden fence. The tall French windows were still there. Even the gas lantern whose glass I had shot out when I was eleven was still atop the pole on the sidewalk in front of the house. As I stared in disbelief, a young man opened a window and called to us across the front yard, asking what we were searching for. He invited us into the house, which was being modified into a residence for working women. My room on the second floor still had the same patterned blue wallpaper as it had when I was born there in 1917. The young German Communist was joined by a teen-age girl; both were very friendly when I told them that I had been born in this house, and that the other members of my party were my American family. They were just going to show us around the old family home when an elderly lady showed up, who blistered when she heard we were Americans—threw us out of the house and suggested we go back where we came from!

Six years later, though, Clarice and I returned once again. This time we drove in a Russian-made LADA, a Hertz rental car, from East Berlin. Frankfurt/Oder had greatly improved. Six high-rise buildings were in town. Stores were fuller; people seemed happier. Clarice and I were invited for *Kaffee und Kuchen* both in the house adjoining my father's factory as well as at our villa. We stayed several days and visited the Polish side of my hometown. (Frankfurt/Oder's western part was given by the Allies to East Germany, the eastern part to Poland.) It's there that I was an apprentice fifty years ago, and an automobile garage is still located there. Yes, you *can* go home again!

In 1981, I received an inquiry from the General Electric Aircraft Engine representative in Peking, asking if my wife and I would accept an invitation from the Technical Ministry of the People's Republic of China to give two lectures before their aerotechnical universities in Peking and Xian (formerly spelled Sian), visit an aircraft engine plant there, and then travel to any place in China we wanted to, especially the combat areas in which I had been active during World War II. If our answer was yes, such an invitation would be forthcoming. Anna Chennault, widow of the General, escorted us to the ambassador of the PRC in Washington, who welcomed us formally.

In October 1981, on a (Communist Chinese-owned) Boeing 747 flight from San Francisco to Shanghai, Clarice and I arrived in the great city on the East China Sea we had last seen thirty-four years earlier. We were accompanied by a charming Chinese-American couple, Walter and Betty Chang, who served as interpreters during our tour through mainland China and my presentations at the two universities. Instead of the international, bustling Shanghai we had left in 1947, before the Communist takeover—crowded with rickshaws, automobiles, trucks, buses—we found a new type of population living in the same old buildings. Old and young people performed physical exercises on street corners and public places at 6 A.M.; workers bicycled to and from jobs by the tens of thousands. The most obvious difference between 1947 and 1981 was the total absence of the masses of coolie-pulled rickshaws and of privately owned cars; store shelves were now nearly empty and the hotel food was bad. But the young Chinese people looked happy, neat and clean; we found an unexpected curiosity and friendliness toward Americans as they crowded around us wherever we stopped. We talked with Chinese students and waiters, bus drivers and housewives, and we found them all just as friendly as their countrymen had been during and after the war under Chiang Kai-shek.

In Peking, an American-educated seventy-year-old vice chairman of the Communist Party hosted us for dinner in the Great Hall of the People. He and I showed up together on Communist national television. The Neumanns did things all tourists do when in Peking: admired the Forbidden City, the Summer and Lama palaces, visited the small section of the Great Wall open to the public. I then talked to an auditorium full of male and female aviation students at Peking University. At Xian we admired the fabulous and incredible excavations of thousands of life-sized foot soldiers, officers and cavalry of the Emperor Qin Shi Huangdi's regime in the third century. I lectured at the aeronautical university in Xian to students and members of the faculty, with the help of interpreter Walter Chang. From Xian we flew to Kunming, the 1941–42 headquarters of the Flying Tigers and the 1942–45 headquarters of the 14th U.S. Army Air Corps. Instead of 150,000 Kunmingnese in Yunnan's capital, there now were over two million! The four beautiful city gates and the city wall had been torn down. Even the picturesque Dianchi lake was being changed through landfill and discharge from heavy industry. Only the two tall pagodas near which I repaired the police chief's Buick were still standing. Modern hotels were under construction. Six- to eight-lane-wide avenues were being built for the future, but currently used only by thousands of bicycles. To my amazement—and thanks to Walter Chang's patience and persistence—we did locate my 1940 home in the Model Village, which used to be outside the city wall of Kunming but which was now well inside the greatly enlarged city. The small European-style houses had been divided into even smaller apartments. Even the Flying Tigers' barracks—originally two miles away from the city wall—were well within new Kunming, modified into apartments.

We flew to beautiful Kweilin on a Russian-built plane. Here took place the center of our fighter activity against the Japanese from 1942 to 1944, and where I had reassembled the Japanese Zero fighter plane after it crashed, following its first flight in 1943. The Kweilin people were most pleasant to us—but made me realize how old I had grown and how restricted Communist education was when I learned that the new generation had not even heard about *any* American flyers ever having been in China during World War II. We decided to look for old, white-bearded men who might still remember the Flying Tigers. True enough—we found two men who did. They advised us through what part of town we should drive (in the government-supplied station wagon with Communist chauffeur) to find our abandoned American fighter strips. Once we reached the former airfield overgrown with high grass and I had a minute to orient myself, I discovered the large stone roller used for flattening the runway after it had been bombed and after the craters had been filled with gravel. I also found the empty command cave of General Chennault and the pillbox in whose gun slot my little Pekinese dog, Tiffy, given to me by a German pilot of Eurasia and which Colonel Chennault let me keep in the AVG, was blasted by a Japanese bomb exploding nearby; he was thrown against the opposite wall with such force that he broke his front leg.

After a few minutes of silent reflection remembering the sunrises, the dogfights overhead, the young Korean woman Angela and my Zero, I was ready to return to downtown Kweilin. An open-bed truck with about twenty young workers standing in the rear and one elderly supervisor sitting in front slowly rumbled by; then the truck stopped. Their leader asked our chauffeur if we were lost. Our interpreter held a brief discussion with him (obviously telling him who we were). The truck drove off. After a few hundred yards it stopped for a minute, then turned back. As the workers drove by us slowly, all of them raised their right thumbs in salute and yelled *"Ding how, ding how!"*—which, as the elderly supervisor must have informed them, was *the* friendly greeting of the Chinese population for our American soldiers in 1942–45. *(Ding how!*—Very good!) I could have cried when I remembered the old days in Kweilin—and the extraordinary, thoughtful gesture of that group of friendly Communist workers which had just passed. To sum up our trip to Communist China: We were most impressed. Though still many, many years behind their brethren in Taiwan, the Communists in China will eventually catch up. They are already freer than the Reds in East Germany, and they will be a power to reckon with in the year 2000 and beyond. After all, aren't they already one of the five nuclear world powers?

We stopped in Tokyo to pay our respects to my Japanese friends at the IHI (Ishikawajima Heavy Industries) aircraft engine plant, which I had helped to reestablish after the war. IHI has built GE engines under license over the past twenty-five years. They invited us to visit the latest modern addition to their steadily enlarging jet engine factory. A Western-style luncheon was served by the general manager Dr. Kaneichiro Imai, who had been installed after his charming and brilliant predecessor, Dr. Osamu Nagano, had retired at seventy. Dr. Imai suggested a walk around the landscaped factory grounds and casually pointed out a tall young tree held upright by supporting wires and taped against bugs and ants. In front of the tree was a large irregular stone with a bronze plaque imbedded in it that looked to me like a gravestone. I paid little attention to it; Clarice pulled me by the sleeve and whispered, "Read it!" Walking closer to it, I noticed the inscription in Japanese and in English: "This Ginkgo Historical Tree planted in commemoration and friendship to Gerhard Neumann, Vice President, General Electric Company."

At about the same time, Jack Parker, to whom I had reported for twenty-one years at GE, flew with his wife, Elaine, to Communist China for a week's visit. After they returned, Jack sent me the photograph he took of a tall white-haired man who, with his wife, had traveled to Shanghai on the same plane with the Parkers. It was a picture of Mr. and Mrs. W. Langhorne Bond! He was the American who had spirited me out of British internment in Hong Kong forty-four years ago.

Old travelers never die—they just keep on flying. In the autumn of 1982, Clarice and I flew around the world once again. We took our time: over two months. We stopped at some of our favorite places in Europe and Asia. One day, fifty miles southeast of East

Berlin, we rented two single-seater kayaks and paddled a few hours in the Spreewald, a forest crossed by many canals of the River Spree. We passed large boats loaded with Communist tourists. I wore a T-shirt with the slogan *"Es ist besser in den Bahamas"* ("It is better in the Bahamas"). A few years ago I would have been locked up by the Nazis or the Communists. But this time the Germans laughed out loud—and some of them even applauded!

20

Looking Back—and Looking Ahead . . .

"Don't ever look back; only look ahead!" we often hear. Obviously, one cannot follow this questionable advice when writing an autobiography. It has been a lot of fun to try to recall, to the best of my ability, some of what happened years ago, to look through scrapbooks and photo albums to refresh my memory. In spite of several close calls I've had in my life, I made it to 1984 and hope to stick around still a bit longer.

There never were dull moments in my life. Some were miserable—but never dull! It made me feel proud to be a member of a superior team like the Flying Tigers, the GOL-1590 engine project, or the General Electric Company itself. I felt especially pleased to see the people my staff and I had identified as "top-notch material" occupy key positions in American industry and do a good job. (I used to play chess or bridge with those whom I interviewed during the lunch hour, to get a pretty fair idea of the person's logical thinking, his carefulness, timidity and aggressive spirit.) Currently, at General Electric, 27 percent of the operating vice presidents are alumni of or still with the Aircraft Engine Group. Hardly any one of these executives attended the company's, or some university's, management training course in a classroom. Instead, he learned to manage in day-to-day operation in my Engine Group, where enthusiasm, leadership, professional know-how and the ability to inspire others are the main criteria to become a manager.

My own career and exciting past were undoubtedly a matter of a great amount of luck, but also of some sweat and tears, often not obvious to the casual onlooker. On our last round-the-world trip a few months ago, Clarice and I stopped overnight at an Austrian village inn where a wooden sign, nailed over its entrance, read: *"Nur mit Arbeit von früh bis spät kann Dir was geraten / Neid sieht nur das Blumenbeet aber nicht den Spaten."* Freely translated it says that only through labor from morning to night can you be successful; jealousy sees only the flowerbed—not the spade.

I am fortunate that most of my personal and business friends, military customers and industry co-workers at all levels—throughout the whole world—were and are decent fellows. Most reflect the highest degree of integrity and ethics. I don't believe that this is a coincidence: The engineering profession, in whatever country, deals mainly with measurable facts, takes technical and financial risks, admits mistakes in judgment or errors in assumptions. I am proud to be an engineer.

In dealing with the military, I found the U.S. Department of Defense nearly always businesslike and fair. Naturally, we faced much more red tape in contracting with the government than with commercial companies. Compensating for the extra paper work, however, was the knowledge that our government was not going to go bankrupt, nor would it demand millions of dollars of concessions in one form or another—as it became, unfortunately, customary in the commercial airline business. Uncle Sam always was fair in assessing penalties for our missing dates or performance; or in handing out incentive awards for our meeting milestones ahead of schedule, producing engines at lower cost, or bettering our contracted performance. President Dwight D. Eisenhower's famous warning to the American people alerting them to the danger of a "military-industrial complex," implying a conspiracy between both parties to build up a bigger war equipment industry, either was not meant by him the way it was interpreted after his death, or the general was not aware of the exhaustive mutual efforts made to improve the equipment of our armed forces, at the lowest possible cost to the taxpayer. While I admittedly cannot talk for the whole defense industry, I have been able to observe many military-industry relationships. All seemed aboveboard and according to the highest business standards. Our government did—and still does—employ a great number of dedicated, motivated military and civilian technicians who receive little or no recognition, and much less pay from Uncle Sam than they would be able to earn in industry. I don't know of a single case where the military services either ordered equipment not thought to be needed or accepted blindly any cost quoted.

During my long career I have seen many talented people remain in relatively low positions in large companies. Breaking through the low-level barrier into a mid-organizational level is the most difficult part of anyone's progress in a big organization. Once a person who has something on the ball has succeeded in penetrating the middle ranks, the chances are good that he will be personally recognized as a contributor. It is then much easier for him to be promoted into still higher positions. In my own case at General Electric, my progress from the very bottom, i.e., test engineer, was not difficult since I was recognized quickly by the jet engine people already aboard as either "that damn foreigner" or the "Red Baron" (Germany's flying ace of World War I, Manfred von Richthofen). Also, I spoke with a German accent, which in the technical arena produced some undeserved advantage over Americans. Sure, I had gone through three years of tough apprenticeship and a practical engineering education which few Americans of my generation had; for that matter, the young engineers in America of today most regrettably still do not have such badly needed training. My career was also helped by my having been fortunate to remain in my professional field of mechanics and aircraft throughout my whole life, in peace and in war. I thus had an opportunity to develop a superior *feel* for technical "go" or "no go" designs in the real world, regardless of theory or computers. Finally, I was continuously lucky. I would have made a good general for Emperor Napoleon Bonaparte, who wrote, "I want my generals to be lucky." (He figured that he would not win many battles with unlucky generals.) Time and time again, I met the right people, at the right time, at the right place. And I was given ample opportunity to succeed—or to fail.

On our last world journey Clarice and I were astonished by the amazing economic advances made in the Oriental countries during the last few years. Old-timers who may recall the Bangkok and Hong Kong of years ago, for example, will hardly believe that the jungle in Thailand is disappearing bit by bit. High-rise buildings and luxury hotels, paved highways, air-conditioned supermarkets and jet plane runways are being built at a rapid rate. There still is an elephant training school in the north of Thailand near the Burmese border, but it is more a tourist attraction than for real, I assume. So are native villages outside Bangkok where Clarice, Chipsy and I stayed in 1947.

The abolition of coolie-pulled rickshaws and of picturesque sailing junks from Hong Kong's scene is sad for photographers, but a sign of progress. Typical four-story-high colonial-style waterfront buildings with sidewalks protected against sun and rain by block-long overhangs have given way to gleaming-white glass and steel high-rise buildings. The great number of Japanese-built automobiles, and of McDonald's or Burger King restaurants in the strangest places of the world, is unbelievable. Tourist groups from all around the world fly into the Orient en masse; television antennas jut skyward from the poorest shacks covered with rusty corrugated tin roofs, in even the most remote parts of southeast Asia; motorbikes by the tens of thousands carry men and women to work. We are slowly and steadily approaching one world—and thus slowly losing the tremendous advantage this country had over the Orient.

Retired now for four years and mostly busy with writing this book, traveling and sailing, I have had time to reflect beyond jet engines. I am very disturbed about some of the peace-at-any-price movements that are developing here and in Europe; movements that are well meant, I am sure, but totally unrealistic and outright dangerous to the Free World. To reduce the Administration-proposed effort of building up strong U.S. defense forces and modern nuclear weapons is currently a popular aim, but in the light of my extensive experiences a totally wrong way to proceed. I am frightened to note how shortsighted and naive even intelligent people—supposedly objective commentators in the media, college professors and some of our Congressmen—are: Already as a boy I had noted how school bullies tackled only smaller kids but never boys their own size. Similarly, I have never seen one country tackle another one when there existed, in the minds of the aggressor, the slightest possibility of his perhaps *not* winning the fracas. The last five decades, which I have personally witnessed, are full of examples from which we *must* learn: Hitler invaded Poland only after he was convinced that his troops would win a quick victory, before the Allies would pull themselves together and live up to the defense obligations they had made to Poland. Hitler's blitz invasion of the west in 1940 took place because he was positive that he would win quickly there, too. Italy declared war on France in 1940 only after France was practically down and out. Russia marched into Finland and Poland only when there was no likelihood of her losing the conflict; the Soviets' occupation of Afghanistan was no different. Some nations tangled with others when they were certain that they would come out on top—but they made mistakes in their evaluations of the enemy. The Argentines invaded the Falkland Islands, never imagining a major British effort to defend them. Look at our USA and the disastrous march into North Korea; our intelligence had failed—and fifteen years later our fight with North Viet Nam. In both cases we had figured that we would win easily.

The following may be news to most activists who advocate the elimination of nuclear weapons: Note that germ, chemical and gas warfare was not introduced in World War II because *both* Germans and Allies knew that each could do the same harm to the other country! Both sides had a full arsenal of these most deadly weapons—as horrible as nuclear ones but not as spectacular. Do you believe for a minute that we would have dropped the atom bomb if we had known that our enemy had such terrible weapons in his arsenal? I am firmly convinced that the *only* way to assure peace between the USA and Russia—whoever may be the man in charge of the Soviets and the Americans at the time—is a strong U.S. defense *equal* to that of the Russians. All of us hope that the weapons we are building will never have to be used and will ultimately be scrapped for obsolescence. Such "deliberate waste of the taxpayer's money" is very much cheaper than a single week of war with the Soviets. The answer to the rhetorical question of the antinuke people is, Yes, we *will* burn if we freeze. Said the other way: *We won't burn if we don't freeze.*

Totally aside, but most important to civilization, have been and will continue to be the valuable spin-offs of high-defense technologies that are put to good use in every citizen's life: jet planes (and how they changed the world's communication), for example. Or take microelectronics, radar, computers, space technology, laser beams, linear accelerators—and many more. These technologies which we take for granted today originated in our defense industry.

Looking back, I cannot help but see another mistaken trend that is developing in this country and in Europe: to put full trust into and to derive the feeling of security from treaties and agreements. Sometimes they *do* work, and I am all in favor of trying to reach agreements on weapon limitations, *but only* if those can and will be monitored without *any* restrictions whatsoever. To rely solely on signed pieces of paper is plain foolishness. Remember British Prime Minister Chamberlain returning from his meeting with Hitler in 1938, joyously calling out, "I believe it is peace in our time"? (Two years later the Luftwaffe leveled the British industrial city of Coventry.) In 1930, all members of the League of Nations at Geneva forswore any future military action against any member nation, and signed reams of agreements to that effect. Three years later, the Fascist leaders of Germany and Italy disclaimed any responsibility for their countries' prior democratic representatives who were authorized to sign such agreements. Germany's delegates simply picked up their briefcases and walked out of Geneva, never to return. Any nonaggression pact signed became totally useless. Instead, Hitler participated openly in Spain's civil war to experiment with some of the new German weapons and military tactics. A few months before, Mussolini drove Emperor Haile Selassie out of Abyssinia and occupied it himself.

Here are more samples of agreements and treaties that were nonenforceable: In 1937 Hitler reaffirmed a ten-year nonaggression pact signed with Poland in 1934, two years later he signed another one with his archenemy Soviet Russia (whom he had blasted for twenty years). He revoked each peace agreement unilaterally. He marched into Poland in 1939 and practically leveled it. He tried to do the same to Russia in 1941—and very nearly succeeded. When Hitler was questioned in 1940 about the ethics of signing peace and armament agreements and then violating them, he replied, "The victor never has to answer questions."

Today, Communist China claims that the ninety-nine-year lease agreement for the southern tip of mainland China, the Hong Kong New Territories, was signed "under duress" with the British over eighty-five years ago and is therefore invalid. The USA and Russia, together with other nations, signed not very long ago a solemn agreement about the exclusively nonaggressive use of outer space. I wonder what is going on in space right now. . . .

Looking back, I am also concerned about the world's sagging respect for America. In the twenties, when I was a schoolboy in Frankfurt, the Americans were our idols: slim, athletic, tall—epitomized by Charles Lindbergh: technically seemingly unbeatable; building powerful and long-lasting automobiles; offering golden opportunities to anyone who wanted to work hard regardless of his background; giving newspaperboys a chance to become millionaires on Wall Street; winning each Olympic Game by a wide margin. We respected America even though the USA, by its intervention in 1917, made Germany lose World War I. To visit America was the dream of each civilized person around the globe. To own an American automobile was heaven. The world's image of America remained high until after Kennedy's death. Then, somehow, our reputation and the regard for us began a slow downhill slide which has not yet stopped. In many social and technical fields other nations have already pulled even or surpassed us. Taking our individual freedom for granted is flirting with ultimate disaster. Many of our younger generation—too young to remember the last big war, what caused it and what miseries and tragedies the world went through—make it abundantly clear that they are not interested in what happened before they were born. The personal freedom we all enjoy is worth fighting for if necessary, a fact young Americans don't seem to realize until they have had a chance to visit a dictatorship.

Another major item on my personal worry list: Are we seriously concerned about the power and influence on world economics that have been brewing in the Orient? Looking back forty-five years, I recall clearly the sleepy, nonindustrial Hong Kong. Today it is pretty difficult to go into any store here in America and not find items marked "Made in Hong Kong." Thirty years ago, I saw flattened areas of destroyed Japan, battered Korea, bombed-out Singapore—and I have watched them rise out of the ashes miraculously. There are also China and Taiwan to reckon with, populated with over a billion industrious people who form a tremendous reservoir of willpower and intelligence. They have an incredible determination to outperform the West—Europe included. People in the Orient are willing to live at a lower standard than we are; and, most important, they are proud of their workmanship. Their governments make long-range plans, and fund the research and development that are necessary to accomplish Oriental supremacy. All this should alert and prepare us for the real possibility of a switch in technological and business leadership in a generation or two, from the West to the East.

The Japanese workers in one major factory, Clarice and I were told in Tokyo, meet once a week after working hours, on their own personal time, to discuss quality problems discovered in their work during the preceding week; they then recommend to their management what improvements in equipment, tooling or production processes are essential to deliver flawless products. Let's look at the fantastic progress made by these people during the last thirty-five years and recognize that the Orient has already taken over leadership from the West or made major inroads in such large-ticket items as automobiles, electronics, shipbuilding, steel production and optics; it doesn't take much imagination to see that the Western nations might have to play second fiddle in the year 2000 and thereafter. Can you imagine what this would do to the style of life we are used to?

Let's change this downhill course! There is no good reason for the American automobile industry not to make every effort to regain its #1 position in the world. (The Detroit people won't agree that they are not #1 now, but the great majority of Americans who have ever driven a recently built foreign car will agree that equally priced foreign automobiles are superior.) Let Detroit look around the world and note how the American car has practically disappeared. Once our car manufacturers make up their minds to apply good, old-fashioned American ingenuity, imagination and design talent to their new car models, once their management decides to invest whatever it takes to translate their new advanced design into a superior quality product, and once the unions agree not deliberately to slow down their workers but encourage them to work harder, there will be no stopping America from reestablishing its leadership in the field of automobiles. Perhaps it'll take one, two or three years of not paying out dividends to shareholders, or not granting salary increases and bonuses to their own management; but funds *must* be made available for engineering, modern production tools and facilities, and a much more effective quality control if we want our private industry to win in the face of the support overseas manufacturers receive from their own governments. If we are not successful, we should get rid of the incumbent managers and replace them with people who aggressively *lead* engineering and manufacturing teams. There *are* people in this country who can succeed in bringing us back into the #1 position of automotive world leadership. And such industrial revival should not be limited to automobiles.

Our government-managed scientists and workers were able to land Americans on the moon and bring them back to earth safely —truly a magnificent engineering feat which still boggles the imagination. Don't let's stop our demonstrated creativeness and leadership.

We have to train our people to work better and inspire them to have more pride in their work. We must be free to select the most capable and imaginative of our managers and workers to do the most important jobs. To be free to do this we will have to modify this country's seniority system, which has been a stone around industry's neck. I know. I lived with it. Today this seniority system is hurting the workers more than management: The large number of unemployed people in America is, in great part, due to the unacceptable quality of many of our products made here. We all would be better off if a supervisor could select the most capable, the most competent, the hardest-working and the smartest of his workers and promote them to handle key jobs, regardless of their seniority. Sooner or later management and union leadership, too, will recognize the problem caused by our current seniority system, which is being handled like a sacred cow. A good part of the present unemployment is the result of inferior American products, their high prices, and the consequent import of foreign replacements. The current seniority system was adequate years ago, when America had unchallenged leadership and no real competition in just about everything. Those days are gone.

Inevitably, I am asked if, knowing what I know now, I would again steer the same course in my postwar career and accept General Electric's offer for the top position in its jet engine business, then hold it for seventeen long years. Climbing the ladder of success was made easy for me and I was rewarded handsomely, not only in regard to the salary I received: There was a lot of recognition, satisfaction and personal pride involved. But the price I paid was very high. The marvelous opportunities that General Electric gave me from the very first to the last day during the thirty-two years I was with the company consumed progressively more and more of my energy as the business grew bigger and bigger. There was barely time left to spend with my wife, who understood the problem and wanted me to do whatever I felt I had to do. I tried to be a reasonably good father, yet my three children had to grow up without much attention from their old man. Every bit of available time was invested in reading the Daily I.O.I.'s covering the business of my Aircraft Engine Group, studying reports, walking the shops. I will not answer here the question, Would you again . . . ? I leave it at that; but I want to alert ambitious go-getters to ponder very carefully the problems that go with accepting a promotion to a top position and the price they will have to pay, before they say, "Yes, sir. Thank you."

The American dream recalled in this autobiography would not be complete if I did not mention a summer day in 1950 when I—with thirty-four others—was called before a judge in the Boston traffic court because of an alleged parking violation. "Guilty or not guilty?" asked his Honor of each driver. A reply of "Guilty!" followed each question until my turn came.

"Not guilty, your Honor!" The judge was surprised, told me to sit down and think it over. After a few minutes and three more "Guilty!" pleas by others, he asked me again. "Not guilty, your Honor!" He doubled my fine. Still, "Not guilty, your Honor. I want to appeal!"

The judge asked me to step up to his side of the bench. He turned toward me and asked, "How long have you been in this country?"

"Two years, your Honor."

He then whispered in my ear that I sounded like quite a decent fellow. "Case dismissed!" he called aloud.

Where else could one read a big black headline in a leading newspaper the following morning: MAN FIGHTS TRAFFIC COURT—AND WINS! Nowhere in the world except America. And I am here! Just lucky, I guess . . .